PILOT ERROR

Anatomies of Aircraft Accidents

By the Editors of *Flying* Magazine

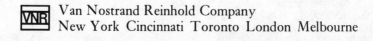
Van Nostrand Reinhold Company
New York Cincinnati Toronto London Melbourne

These accounts are based on actual National Transportation Safety Board accident reports and are intended for educational purposes only. They do not reflect on the ability or capacity of any person, living or dead, or upon any aircraft or accessory.

Copyright © 1977 by Van Nostrand Reinhold Company Inc.
Library of Congress Catalog Card Number 77-3144
ISBN 0-442-22481-8

Published in 1977 by Van Nostrand Reinhold Company Inc.
135 West 50th Street, New York, N.Y. 10020

Van Nostrand Reinhold Australia Pty. Limited
480 Latrobe Street, Melbourne, Victoria 3000, Australia

Van Nostrand Reinhold Company Limited
Molly Millars Lane, Wokingham, Berkshire RG11 2PY England

16 15 14 13 12 11 10 9 8 7 6 5

Library of Congress Cataloging in Publication Data
Main entry under title:

Pilot error.

   Includes index.
   1. Aeronautics—Accidents. I. Flying.
TL553.5.P55   629.132'52   77-3144
ISBN 0-442-22481-8

CONTENTS

FOREWORD

The twenty-five accident reports in this book originally appeared in *Flying* Magazine as a bimonthly column entitled "Pilot Error." The purpose of the column was to allow pilots to learn from the mistakes and misfortunes of others and to avoid repeating them. At the same time the reports were widely felt to offer some of the most gripping reading in the magazine.

The title "Pilot Error" has led to misunderstandings. It is a ready-made phrase commonly used in accident reports to cover all possible contributing factors that in some way or other involve the acts or omissions of the pilot. It does not have a precise meaning. In fact, much of what in retrospect appears as an "error" may at the time have seemed the best of several bad alternatives, and not every move that leads eventually to an accident was necessarily and obviously an "error." Furthermore, the fatal outcome of these cases produces the impression that the "errors" of the pilots must have been particularly grave or flagrant. This is not the case. In fact, what makes these reports especially chilling reading for pilots is the knowledge that they themselves have again and again made the same errors but have in one way or another escaped the ultimate retribution. Finally, the commentaries by William L. Traylor, Richard L. Collins, and Peter Garrison that accompany these transcripts are largely hypothetical reconstructions of real events; assignments of blame are necessarily uncertain and have been avoided unless they seemed necessary to clarify the useful lesson of the story. It is not our purpose to say that this or that pilot acted well or badly: our purpose is to show the ways in which events unfold, again and again, in aviation accidents so that other pilots may recognize and avoid the pitfalls that have claimed so many lives in the past.

ROBERT B. PARKE
PUBLISHER, *Flying* MAGAZINE

INTRODUCTION

Total safety of flight is no doubt an unattainable goal. The aerial environment is a hostile one to man; we do not, when dropped, settle gently to earth like dry leaves. To remain aloft requires the cooperation of many elements, human and mechanical; it is too much to expect that, over the course of dozens of years and millions of flights, one or another of those elements will not from time to time fail us. Nevertheless, retrospective examination of aircraft accidents reveals that many of them occur unnecessarily. The element that failed *was* within our control. One class of such needless mishaps is the accident that is brought about or contributed to by the misjudgments or gambles of pilots. Almost all the accidents in this book are of this type.

This is not a typical cross section of accidents. For one thing, it is heavily biased toward IFR accidents, because the format of the articles required a transcript of recorded radio communications or cockpit conversations as a starting point. Obviously, such transcripts are available in a minority of cases but most commonly when the doomed aircraft was on an IFR flight. The "pilot errors" depicted here are many, ranging from serious misjudgments of weather conditions by inexperienced pilots to incorrect assumptions about their position by highly experienced airline captains. Two factors contribute to many accidents but receive comparatively little attention in accident reports: one is the relationship between the pilot and the controller; the other is fear.

An axiom, by now a cliché of aviation, is that the pilot has the final responsibility for the safety of his aircraft and passengers. He is the final authority in all decisions. The position is a traditional one, handed down to aviation, no doubt, from shipping, along with many other expressions and attitudes. The pilot *is* the flight; he invests it with its identity and character, rather as the captain does a ship. The analogy between the ships of the air and those of the sea is very rough. Many functions of command in the air are obviously cooperative: the pilot and the ground controller share responsibility for the flight, and the pilot's authority is only figuratively "final." Practically, of course, it *is* final, because he holds the yoke and the throttle in his hands. But he is helpless, despite his authority: without the controller, he would not know what to do.

Because the functions of pilot and controller are so obviously cooperative, and because, whatever his theoretical prerogatives, the pilot daily finds himself obeying the controller's orders, the controller has a psychological advantage over the pilot. Under IFR procedures, the pilot is often a mere technician executing heading and altitude commands, while the omniscient controller rules the skies.

In general, this system works satisfactorily. There have been rare cases of controllers' errors leading to accidents, and presumably there have been errors that have gone unnoticed or occurred without serious consequences. Pilots' errors are generally more prolific of accidents. But what several of the accounts in this book call attention to is the precarious situation that exists when the pilot is unable to assure the safety of the flight and must hand over some portion of his control or all of it to the controller.

Most commonly, it is the relatively inexperienced pilot flying into difficult weather conditions who ends up abdicating in favor of the controller. Sometimes it is a pilot who is lost, though in good weather conditions. In either case the pilot feels helpless—in fact is usually powerless to extricate himself—and the controller appears to be his only hope. Though it is in principle the pilot, not the controller, who retains responsibility for the flight, it is in fact the controller who is in command, though he may not wish to be and must always have in the back of his mind a sense that what is happening is not what is supposed to be happening.

The dilemma arises from the nature of flight instruction and the attitudes it instills. Very few pilots regard flying as just another activity, like driving a car or roller skating. They feel that there is something special about it, engendered by the life-and-death seriousness of being airborne. Flight is regulated in an authoritarian, paternalistic way, the Federal Aviation Administration wielding an unassailable moral authority as the protector of life. Instructors lord it over students, high-time pilots over low-time pilots, experts over novices; minor mistakes and omissions are strongly censured as being potentially productive of gory disasters. Shame and humiliation are visited upon the pilot who violates the high standards of the fraternity.

Paradoxically, many highly experienced pilots pride themselves on their enlightened indifference to the rules; they are often inclined to view the makers of rules as drones cut off from reality and to see themselves as equal to all situations on terms of their own making. It is notorious on the "inside," for instance, that the fact that so many airline flights are more or less on time is due to the willingness of so many airline pilots to descend below minimums as a matter of course. But to the inexperienced pilot, this side of things is not well known: like a child growing up in a household of Puritan fundamentalists, he is likely to feel that everyone else is better than he and that his own failings are to be hidden at any cost. For males, learning to fly is often tied up closely with their own feelings of identity and masculinity, and the anxiety of guilt arising from the sense of their own shortcomings is deep-rooted and strong.

The disembodied voices of controllers are, to the learning pilot, something like the voice of the Lord speaking from the mountaintop. Few beginners practicing in the pattern escape mortification at the more careful diction that the controller uses for student pilots or at any error or misspeech or hesitation on their own part. They gradually gain in confidence and eventually may regard mere tower controllers as superfluous nuisances, but there is always some higher sort of controller who inherits the fasces and, figuratively at least, the "sneer of cold command."

With the mythology of the controller in mind, it is easy to see how the "pilot in command" may automatically relinquish his command to the authority on the radio, even though that authority may not want the command and be unable to take it. In the case of the young woman lost in the Bahamas (Chapter 13—"Please Keep Talking to Me . . . ,") the parental role of the controller is particularly obvious. Perhaps because she is a woman and less trained than a man in the requirements of machismo, she readily expressess her feelings: "Please keep talking to me . . . I need it." Other pilots may not express the need so clearly, but it is revealed by the constant stream of superfluous communications evidently intended to keep alive the sense of a guiding hand.

In the case of the lost pilot in the Bahamas, the fatal and mysterious outcome—shades of the Bermuda Triangle—makes it clear in retrospect that she made a mistake in putting herself into the hands of the controller. She should have weighed the situation and concluded that a landing on unsuitable terrain on one of the deserted islands that she could see would be her best bet. But because the controller sounded as though he knew what he was doing and, even more importantly, because he was a *controller* (not in the legal sense but in the pilot's own sense of his role), she could not give up the comfort of his assistance. The temptation to shift the responsibility in a hopeless situation is very strong.

In the approach situation, controllers not only have knowledge of traffic that pilots cannot have but also of terrain. Minimum vectoring altitudes, normally known only to controllers and not to pilots, are lower than minimum en-route altitudes and may give pilots the impression that descent below the MEA during an approach is normal and expected—whence the recurrent accidents resulting from premature descent, one of which (Chapter 15—"Use Any Approach You Like") appears in this collection.

In "Use Any Approach You Like" the possibility arises that the pilot felt uncertain of when to begin his descent from cruising altitude and thus called the controller to report his altitude after he had already been descending for some time and was 1,000 feet below the minimum en-route altitude. The report was uncalled-for. Whether or not uncertainty motivated this particular call, it plausibly might have, since all instrument pilots have known the experience of feeling uncertain about a procedure and of not wishing to admit their perplexity to the controller. The regulations regarding pilot qualification for instrument flight seem to boil down to requiring that the pilot know everything and make no mistakes. Pilots, especially inexperienced ones, feel chronically under-qualified for the reasons already described. Each pilot who encounters uncertainty during an instrument flight feels that he is the only incompetent in the sky and that to make a radio call admitting his doubt would not only broadcast his failure but also probably bring some FAA official out to greet him at landing and perhaps relieve him of his license. That scenario is quite implausible, but it is the sort of thing that children and pilots fear. The pilot in this case was a flight instructor and probably felt a particularly strong aversion to admitting uncertainty or ignorance, as doctors often do.

Uncertainty brews anxiety, anxiety swells into fear, and sometimes fear explodes into panic. For the novice instrument pilot—and the novitiate may last for years—this is a most dangerous sequence of events. For the non-instrument-rated pilot who wanders into instrument conditions the sequence is a virtual certainty. The most vivid, almost grotesque example in this collection is "Panic" (Chapter 6), in which a qualified but anxious pilot is rendered incompetent by an insignificant instrument malfunction.

Fear is not dealt with at all in instrument instruction. At the very early levels of flight instruction it is swept under the carpet (". . . you see, there's nothing to be afraid of . . ."), and there it remains. The pilot is conditioned to believe that, if he proceeds according to the rules and is competent, nothing will go wrong. When something does go wrong, he learns, it can always be traced to some error on the part of the pilot. For instance, if the engine quits, it is probably because of improper preflight or inadequate maintenance. Again and again, pilots are found to show little sympathy for their colleagues who are hurt or die; some simplifying explanation is immediately hit upon to reassure the others that the same fate will not be theirs. They conspire to ignore and forget that "no man is an island" and that, when the bell tolls for another, it tolls for them as well. The telltale exception is that pilots like to tell stories of near disaster in which they may mention in terms of comical exaggeration how scared they were (". . . about to wet my pants . . ."), but the nonchalance of the description is meant to produce the appearance that they have sloughed off this fear or were able to deal with it coolly. Fear is the leprosy of aviation.

Because of the instructional situation, in which the student always feel able to rely on the instructor to save him from his own blunders, the worst fear that arises is fear of failure or of loss of face. It is trivial compared to the fear of loss of life.

In several of the cases in this collection, the pilot is face-to-face with his own death. The fate of his passengers may or may not absorb him, but at any rate he thinks, rightly or wrongly, that he is about to be killed. Why this frightens people would be an interesting subject for discussion, but for the moment, it is sufficient to agree that most people find death about as frightening as anything can be. Some of the pilots handle the pressure well—generally, the more experienced "high-time" pilots, professionals, who have learned from long exposure to fear how to control it. The pilots in "Unable to Maintain" (Chapter 4) and in "Can't Continue on that Left Engine" (Chapter 20) are in the worst imaginable situation: airplanes crippled, unable to climb, sinking into the terrain. Yet they remain in control of themselves, though their guts must be in a turmoil. It is the inexperienced pilots, confronting the unknown for the first time and deprived of the comforting sight of the ground outside the windows, who lose their grip in case after case. This must seem puzzling to people who have never experienced or have forgotten the cold and alien stare of gray windows, the comfortless darkness, and the clogging of the mental processes that at once makes it seem desperately urgent to turn to a particular heading and impossible to decipher the number on the directional gyro. It is almost impossible to imagine a situation in which one would be unable to interpret the reading on an omnibearing selector or unable to tell if one is turning left or right, and so these quite real eventualities are not even mentioned in training. Students might be discouraged from going on if they were told that they might be in training for the most awful moment of their lives. Yet these are precisely the moments that take lives. Eventually, most pilots come face-to-face with their moment of stark terror. Most of them return alive, and they have learned things about flight and about themselves that instructors could never have taught them and that the *Airman's Information Manual* never mentions. They are the ones who, years later, may be able to face the ultimate danger coolly so that novices seeing them will in their turn imagine that those old-timers never knew fear.

In *Night Flight*, Saint-Exupéry describes an airmail pilot who has flown into a storm at night and gotten thoroughly frightened. Seeing that this fright may sow a cancerous seed, Rivière, the line's coldly paternal dispatcher, dismisses the incident, telling the pilot that there was nothing out there to fear—nothing that was not there in the clear daylight—but that it was his imagination that frightened him. He teaches him to ignore and despise his fear. Another pilot, the protagonist of the book, is also lost in a thunderstorm at night, low on fuel, and blown out over the sea and sees that he is doomed. He climbs, breaks out on top, and is greeted by the heartbreaking beauty of the moonlight upon the cloud tops that will soon swallow him; he sees all this and his approaching fate with equanimity. He has already learned the lesson and is putting it to use; it does not save his life, but it ennobles him.

Ennoblement is fine for literary purposes but not much help to a pilot about to slam into a mountain. The significance of the events in *Night Flight* is that they show how a pilot must learn to cope with fear and that he can learn only by experiencing and confronting it. The accident reports that you are about to read show again and again how fear is not merely, so to speak, the musical accompaniment to disaster: it often plays a role in producing the fatal outcome.

To conquer fear will not eliminate accidents: the last two cases in this collection involve seasoned pilots who fearlessly, even recklessly overstep the bounds of caution. But fear, like some structural flaw in the airplane, is a contributing cause in accidents that can be eliminated or minimized once it is recognized and confronted. The misguided heroism of pilots, their reluctance to admit that they have felt fear that was sickening, unmanning, humiliating, is a disservice to others who will look to them for example and instruction. Of the many lessons that these accidents contain, the lessons about fear and about the weakness and fallibility of all humans, pilots included, are among the most valuable.

PETER GARRISON

1.

ONLY FIFTEEN MINUTES

On the afternoon of August 16, 1969 a 33-year-old psychiatric technician and one passenger boarded a Piper Cherokee 140 and left Sonoma Skypark at Sonoma, California. The purpose of the flight was to attend a meeting at Compton, near Los Angeles. There was no record that the pilot requested or received a weather briefing, and no flight plan was filed. At approximately 1900 that evening, the aircraft—N9695W—arrived at the Compton Airport, where a friend met the pair and drove them to the meeting. During the course of the evening's activities, it was noted that neither the pilot nor his passenger drank any alcoholic beverages.

At 2350, the two of them returned to the airport, where the aircraft had been topped off with 38.5 gallons of fuel. The friend who had driven the pilot to and from the meeting was then given a 15-minute sightseeing flight of the city, and, upon landing, the original passenger reboarded, and the pair immediately departed on the return leg. Again, no flight plan was filed, and no weather briefing was obtained. The weather over Los Angeles was clear, with good visibility.

There is no record of any radio contact with the flight until 0404, when the pilot called San Francisco Radio for assistance. Santa Barbara Radio responded, and a DF position indicated that N9695W was on a bearing of 260 degrees from Santa Barbara. The pilot stated that he was unsure of his position and gave his altitude as 6,500 feet, heading 130 degrees. Santa Barbara suggested that he fly an easterly course and then alerted Los Angeles ARTCC. At 0434, Los Angeles established radar contact and placed the aircraft 60 miles southwest of the San Luis Obispo Vortac—a position more than 50 miles offshore. The following is a transcript of the relevant radio contacts (95W = Cherokee N9695W; LAX = Los Angeles Radar Control; PRB = Paso Robles Flight Service Station; VBG = Vandenberg AFB's GCA; CGR = U.S. Coast Guard at Long Beach; * denotes contacts between ground stations on direct land line).

TIME: 0434
LAX—Cherokee 95 Whiskey, radar contact, your position 60 miles southwest of San Luis Obispo. Now turn left heading 040 for a vector to Paso Robles.
95W—Zero four zero to Paso Robles.
LAX—Roger, what's your fuel state now?
95W—My right tank is empty, sir, and my left tank is (unintelligible).
LAX—Roger, Cherokee 95 Whiskey, what is your power setting?

95W—Twenty-three hundred.
LAX—Roger, do you have it leaned out?
95W—Negative right now.
LAX—Cherokee 95 Whiskey, roger, uh, lean it out so you can conserve as much fuel as you can.
95W—Roger, I've got it leaned out now.
LAX—Roger.
*LAX—Paso Robles?
*PRB—Paso Robles.
*LAX—How about Oceano and San Luis Obispo—do they have lights?
*PRB—Oh, that's all clobbered in, uh.
*LAX—Oh, it is?
*PRB—Yeah, the only place that's open is Paso Robles, and, uh, unless you steer him out into the valley there, San Joaquin Valley, but, uh.
*LAX—No (unintelligible).
*PRB—Hasn't got enough fuel to get there, has he?
*LAX—No, I don't know if we've got enough fuel to get to Paso Robles.
*PRB—By golly, uh, the whole coast there is indefinite 100 at Vandenberg; Santa Maria should be clobbered.
*LAX—Santa Maria is, but I was wondering about, uh, San Luis Obispo. It probably would be too.
*PRB—Yeah, that would be too.

TIME: 0440
LAX—Cherokee 95 Whiskey, are you by yourself?
95W —Negative. I have one passenger.
LAX—What time did you take off, and did you take off with full tanks?
95W—Took off with full tanks at about 1130.
LAX—Cherokee 95 Whiskey, your position is now 50 miles southwest of San Luis Obispo. Turn right heading 050.
95W—Ten-four.
LAX—Cherokee 95 Whiskey, do you have any instrument training at all?
95W—I have some—about six to eight hours.
LAX—Roger.

TIME: 0444
*LAX—Vandenberg?
*VBG—Yeah, on his heading to Paso Robles—it's almost due north, isn't it?
*LAX—No, 050.
*LAX—Right now from Vandenberg, let me getcha a sweep here, he's about due west of Vandenberg at . . . about 40, let's see . . . 25, 35, 42 miles.
*VBG—Forty-two miles? What's his altitude?
*LAX—Eighty-five hundred.
*VBG—I ain't paintin' a dad-gum thing out there.
*LAX—Uh, he's uh, 050, heading right for Paso Robles, but, uh.
*VBG—Wait a minute, wait a minute, here comes something, yeah, just.
*LAX—Is he on the edge of your scope?

*VBG—Just starting to paint him up . . . gettin', ah, startin' to get a strong target on him, I'll betcha that's him, bearing about 260 from Vandenberg.
*LAX—(unintelligible), yeah.
*LAX—Cherokee 95 Whiskey, your position now is 38 miles southwest of San Luis Obispo.
95W—Ten-four.
*LAX—Coast Guard?
*CGR—Yeah.
*LAX—Okay, position now is from Vandenberg VOR, due west at 40 miles.
*CGR—He'll pass north of San Luis Obispo?
*LAX—Right.
*CGR—I see, and he's en route, or he's heading for San, uh, Robles?
*LAX—He's heading for Paso Robles if he can get there. But his fuel state, I doubt if he can make it.
*CGR—Uh huh.
*LAX—In fact, I don't know whether we'll be able to get him to the shoreline or not. Right now I got him in radar.
*CGR—How far inland is Paso Robles?
*LAX—Uh, be another 15 miles inland.
*CGR—I see.
*LAX—But he's not instrument-rated and he can't down anyplace on the coast. Vandenberg is closest, but there's no way he can get down through it. They've only got an eighth of a mile.

TIME: 0450
*CGR—Could you give us an altitude, please?
*LAX—Just a second.
LAX—Cherokee 95 Whiskey, you still at 8,500?
95W—Negative—I've dropped down to 6,100 now.
LAX—Roger, can you maintain that altitude?
95W—Ten-four.
*LAX—He's 6,100 feet.
*CGR—6,100, okay.
*LAX—Uh, hold on, let me see if he can see this ship down below him.
LAX—Cherokee 95 Whiskey, the Paso Robles altimeter is 29.79, and can you see a ship below you?
95W—I don't see anything below me yet.
LAX—Roger.
*CGR—How far is he from San Luis Obispo now?
*LAX—He's, uh, just a second, 27 miles southwest of San Luis Obispo.
*CGR—Two-ten from San Luis Obispo VOR, huh?
*LAX—The VOR now, not the town.

TIME: 0453
LAX—Cherokee 95 Whiskey, now 23 miles southwest of San Luis Obispo.
95W—My left tank just went dry; I switched over now, and I'm on my (unintelligible) right tank.

LAX—Roger, the uh, left one didn't run dry yet, did it?

95W—Ten-four, it just went dry, and I switched tanks and, uh, just a little bit in the right tank.

*CGR—Uh, what's his course now?

*LAX—Heading 050.

*LAX—Okay, he just ran out of gas in the right tank or left tank, and he switched over to the right tank, which shows empty on the gauge, so we ain't got much time.

*CGR—Okay, what's his true airspeed?

LAX—Cherokee 95 Whiskey, what is your airspeed?

95W—Eighty-five, right now.

*LAX—Eighty-five knots.

*CGR—Okay, uh, let's see what we can get out of Point Mugu.

*LAX—You got an aircraft available?

*CGR—Uh, not out of Point Mugu, I mean nothing, uh, Coast Guard has nothing up in that area. What's his position now?

*LAX—He's, uh, on a 210 radial, he's 24 miles southwest of San Luis Obispo— that'll put him about 20 miles off the coast.

TIME: 0457

LAX—Cherokee 95 Whiskey, you're 18 miles from the shoreline now.

95W—Ten-four, I can see lights in front of me off to my left and (unintelligible) off to my right.

LAX—Roger, are they at a distance about 10 or 12 miles away?

95W—Ten-four.

*LAX—Paso Robles?

*PRB—Paso Robles.

*LAX—Yeah, uh, he can see the coastline, uh, can see San Luis Obispo; is there lights available there, lights on the airport?

*PRB—Well, there always is, yes.

*LAX—Okay and, uh, how about Oceano? Do they leave their lights on?

*PRB—They should have, but if he can see it, fine and dandy, but, uh, generally it's under fog.

*LAX—Yeah, well the thing is, we don't want to go by generally. If there's a chance that it might be open, uh, but there's nobody out at, uh, San Luis, is that right?

*PRB—There's no one, uh, that's not manned right now, uh, there won't be any until early in the morning about six o'clock.

*LAX—Okay.

LAX—Cherokee 95 Whiskey, you still level at 6,000?

95W—I'm back down to 5,300 now.

TIME: 0459

LAX—Roger, just hold that—don't start a climb.

95W—Ten-four.

LAX—Can you see the coastline straight ahead of you? Is that correct?

95W—I can see some lights.

LAX—Cherokee 95 Whiskey, can you see a rotating beacon? Look out about between one and two o'clock to your position and see if you can see a beacon. A green-and-white rotating beacon.

95W—Negative right now.

LAX—Roger, keep looking in that area. Let me know if you can see a beacon at all.

95W—Ten-four.

LAX—Cherokee 95 Whiskey, you're 17 miles from San Luis Obispo VOR and about 12 miles off the coastline.

95W—Ten-four.

TIME: 0502

95W—I can see one white light, but it's not rotating, about 11 o'clock.

LAX—Cherokee 95 Whiskey, roger, that would be, uh, Point San Luis Lighthouse. Keep your eyes to the other side of it and see if you can see a rotating beacon.

95W—Is it just back of the lighthouse?

LAX—Just beyond the lighthouse about five miles you should be able to pick up a rotating beacon if it isn't obscured by fog.

95W—I can't pick it up yet.

TIME: 0504

*LAX—Vandenberg?

*VBG—Yeah, did he turn north toward San Luis Obispo now?

*LAX—Yeah, I got him going out there at San Luis now.

*VBG—Yeah.

*LAX—We'll see if he can see the airport as he goes by.

*VBG—Uh huh.

*LAX—But, if not, we're going to have to go to Paso Robles.

*VBG—Good luck, Charlie.

LAX—Cherokee 95 Whiskey, can you still see the lighthouse?

95W—I can see the lighthouse and a radio station, er, radio beacon . . . it's just a radio beacon because it looks like it has some lights up and down on it that are flashing.

LAX—Cherokee 95 Whiskey, Sheriff's Department advises that the San Luis Obispo Airport is under fog right now, and, uh, we're going to just continue on by it toward Paso Robles, and, as soon as you get sight of the ground, advise us, please.

95W—It looks like I can see ground between me and that radio station. I don't know if it is or not, but that's what it looks like.

LAX—You can see ground between you and the radio station?

95W—It looks like it, and it looks like there's water or something on the other side. I'm not positive. I just ran out of gas!

TIME: 0506

LAX—Cherokee 95 Whiskey, you're five miles southwest of the shoreline. If you can just glide in as close to shore as you can get . . . Cherokee 95 Whiskey, try the other tank now and rock your airplane from side to side.

LAX—Cherokee 95 Whiskey, turn on your electric fuel pump; put your mixture to full rich.

95W—I have just switched tanks, and it's in a little bit but not much.

LAX—Roger, back to the other tank and rock your wings. Full rich with your electric pump on and see if you can get a little more out of it.

LAX—Cherokee 95 Whiskey, if you can still see that strip of land that you had, uh, your eye on, try to get to it.

95W—That was straight ahead.

LAX—Roger, you're pretty close to the shoreline now; you're about four miles offshore.

LAX—Cherokee 95 Whiskey, the elevation of the ground in your direct path is 1,800 feet.

95W—Well, I'm at 1,500, er . . . 5,000 now, and I'm dropping fast.

LAX—Roger, can you see the . . . can you see the surface at all?

95W—I can see the surface (unintelligible) than anything else.

LAX—Say again what it looks like again.

95W—It looks more like (unintelligible: a ridge) than anything else . . . is there (unintelligible: a big set of reefs down here)?

TIME: 0509

LAX—Roger, you're crossing . . . crossing onto the shoreline now.

95W—The shoreline is?

LAX—Roger, you're on the shoreline now. Might be best if you could make a left turn to about, uh (unintelligible) 020 and parallel to the shoreline and land on the beach.

95W—Ten-four.

LAX—Cherokee 95 Whiskey, try 85 knots and full flaps.

95W—I don't know if it's fog up here or what it is, but I can see the tops here—I don't know if it's hills or what.

LAX—Roger, if you can reach something that you can see and land on, that'll be your best bet. Can you reach the hills and land on top of one of the hills?

95W—Probably.

TIME: 0512

LAX—Cherokee 95 Whiskey, how do you hear me?

LAX—Cherokee 95 Whiskey, do you hear me?

When no further contact could be made with the flight, a search was begun. Shortly after dawn, a U.S. Coast Guard helicopter located the wreckage of N9695W about eight miles from the San Luis Obispo Airport. The aircraft had hit the south slope of a 1,300-foot-high ridge about 100 feet from the top.

Investigation at the scene revealed that the fuel cells were empty and that there was no evidence of fuel spillage. The engine cowl was destroyed, and severe damage was evident rearward to the instrument panel; the aircraft remained basically intact and had settled on top of a dense growth of underbrush. The wings, main landing gear, aft fuselage, and empennage were not damaged. The pilot's seat belt had failed due to impact stresses, and both occupants were killed in the crash.

The tragedy that stalked Cherokee N9695W across the fog-shrouded coast of California and out over the dark Pacific was relentlessly encouraged by a continuing series of minor errors. By neglecting even the most basic rules of air safety, the pilot and his hapless passenger found themselves in a situation from which even the vast resources of the air-traffic-control system and the U.S. Coast Guard could not extricate them.

The first of many mistakes was committed when the pilot set out from Sonoma Skypark on the initial leg of the flight. He chose not to check the en-route weather or file a flight plan at this time, and the fact that he encountered perfect weather along the entire 410-mile course may have encouraged him to repeat this omission of good operating practice on the midnight return flight. The only preflight planning discovered for the trip consisted of five VOR stations listed on a piece of scratch paper, showing the frequencies, radials, and distances between these widely separated checkpoints. The paper was labeled "down," and the course indicated the magnetic headings for the southeast-bound leg to Compton. There was no evidence that the return course had been even briefly computed, since there were no notations of the reciprocal headings on the paper. None of the sectional charts aboard the aircraft was marked with the course of the flight.

If the pilot intended to follow the same route on the return leg of the trip, the course led parallel to the coast from Compton direct to the Ventura and Gaviota VORs. Gaviota is located about 20 miles west of Santa Barbara, a point approximately 120 miles from Compton. From takeoff, on the first portion of the flight the pilot would follow a heading of about 270 degrees to Gaviota. Then the course changes to a heading of 306 degrees to San Luis Obispo, the third of the VORs listed on the paper.

Although the California coast follows nearly a constant magnetic heading of 300 degrees from Los Angeles to north of San Francisco, there is one important exception to this fact. Immediately north of Los Angeles the shoreline bends sharply west, and the coast in this area runs almost exactly east and west. When the pilot left Compton, he assumed the westerly heading of the Ventura and Gaviota VORs, which also led along the well-populated resort beaches. At the time of their departure Ventura was reporting clear sky with stratus layers offshore, while Santa Barbara had a low, broken ceiling with a heavy fog bank at the shoreline. Other stations farther up the coast were reporting heavy fog.

November 9695W then passed over the well-illuminated coastline during the first hour of flight, and the pilot was settled on his general course. As they passed Santa Barbara, the last major city before San Luis Obispo, there would be little ground reference to guide the course of the flight. At Gaviota, where a heading change of 36 degrees was required, the route passes over sparsely populated farms and mountainous terrain. As the pilot passed the Gaviota VOR, he evidently did not notice the course change needed to reach San Luis Obispo and simply continued on his present heading confident that it still followed the shoreline as it had for more than an hour. Instead he soon crossed the shore and headed directly to sea.

More than two hours had passed before the pilot finally realized that all was not well and called San Francisco Radio for help. If he had been on course, N9695W would have been approaching its destination at this time. The fact that the pilot called San Francisco is evidence that he was unaware of his present position, which was some 300 miles from that station. At the time that radio contact was made, the pilot reported his heading as 130 degrees, a course that would not have led him back to shore. Although it will never be known exactly what course the flight followed, it can be reasoned that, after passing out to sea, the pilot probably wandered for a considerable time on an indefinite heading. He may have sighted the occasional lights of the many ships moving along the coastal shipping lanes beyond the fog belt. Thinking these to be ground lights, he would be led only farther offshore. Finally the pilot assumed a heading of 130 degrees, expecting it to lead back to the Los Angeles area, when in fact he was already too far at sea to do anything more than barely reach the shoreline.

Every assistance possible was rendered the stricken flight by the ground controller. The pilot was reminded to lean the fuel mixture; if this had been done earlier, the outcome might have been far different. As the aircraft approached the final moments of flight, with a dead engine and gliding over the coastline in an inevitable descent toward the fog below, the controller advised him to turn to a heading that would parallel the beach. Instead, the terrified pilot chose to continue into known rising terrain in hopes of landing on a mountain ridge rising up through the fog, an apparition possibly more clearly defined by urgent hope than by actual sight. The pilot was even told to extend the flaps as a final means to slow the inevitable impact; the flaps were found fully retracted at the accident site.

Although the individual number of small errors committed by the pilot approaches monumental proportions and the several simple procedures that were not followed include the most basic steps in flight planning, one cannot overlook the fact that the 15 minutes of fuel consumed during that short scenic flight offered to a friend would have taken the aircraft the last 15 miles to the VFR field at Paso Robles.

The final evidence in an indictment of carelessness and neglect was provided in the official accident report. The pilot had worked a full day prior to the trip to Compton. At the time of departure on the return flight he had been awake for over 16 hours. His total logged flight time indicated just over 100 hours, only four of which were flown at night. Found in the wreckage was a pill bottle with the pilot's name on it. The bottle contained tranquilizers of several different strengths, a pain-killer, and some commercially marketed stimulant tablets of caffeine. In spite of all these facts, it must be remembered that this accident might have been prevented by the simple expediency of a line drawn on a sectional chart—one of the first skills learned by a student pilot.—W. L. T.

What is particularly puzzling about the history of Cherokee 95W is the extraordinary length of time during which the pilot cannot have had any radio verification of his position or, for that matter, any other verification. What was he thinking about between the time when he passed out to sea over a stratus layer and the time, over two hours later, when he finally called for help? Pilots who have experienced the sensation of finding themselves lost in the sky, particularly at night, remember how each minute seemed like an hour and how one is gripped by an obsessive desire to get back to any known point. That the pilot of 95W flew out to sea for so long without correcting his mistake is incomprehensible.

The fact that he used up the fuel that might have saved his life in a sightseeing flight is merely ironic: it has nothing to do with the accident, since, when he left Compton, he still had plenty of fuel to reach Sonoma. His choice of a route up the coast was ill-advised, both because of the weather, of which he was apparently unaware, and because the inland route through the San Joaquin Valley is shorter, has better weather, and offers better terrain for an emergency landing.

The controller who tried to help 95W showed excellent judgment in advising the pilot to lean his mixture and to use flaps for his forced landing; he did all he could. If he had known that the plane would run out of fuel where it did, he might have vectored it toward an area of marsh and sand dunes a little south of San Luis Obispo, which might have offered a better place to crash, but there was no way to know how much fuel was left, and a slim chance of making Paso Robles might have seemed better than a certainty of crashing blind in fog.

In this accident, as in many others, the cabin remained more or less intact, and the occupants were killed when they were thrown against the instrument panel by the deceleration of impact. Their seat belts failed; FAA requirements for seat belts include a forward acceleration of only nine Gs. It has been shown that the human body can tolerate 40 Gs if properly restrained. Apart from the many errors made by the pilot— and they are errors made daily by inexperienced pilots without capital punishment—the direct cause of the deaths was quite probably the lack of belts of adequate strength. Again and again, it will be seen that their presence would make the difference between death and survival. The pilot of 95W was inexperienced; he undertook a moderately demanding flight and handled it badly. Other pilots have fared as badly in spite of years of experience.—P. G.

2.

A WINTER'S TALE

A familiar midwinter-weather pattern was developing along the Pacific Coast on the morning of January 11, 1970. The weather surface charts displayed in the U.S. Weather Bureau office in Fresno, California showed a moderate cold front approaching at 20 to 25 knots. By midmorning extensive areas of low clouds and precipitation were covering most of the state, with active clouds and heavy rain well in advance of the actual front. The timbered slopes of the rugged Sierra Nevada Mountains were already obscured by heavy clouds that formed layers up to 12,000 feet. The peaks of granite jutting above 12,000 feet were entirely covered as the layers merged into solid gray at the higher levels. Snow had begun to fall in scattered showers that were gradually gaining in frequency and intensity.

Forecasts included numerous pilot reports from scheduled airline flights moving along the victor airway connecting Reno and San Francisco. Comments about icing in the clouds were frequent. The freezing level had dropped below 9,000 feet and few, if any, light aircraft were venturing far from home field.

At approximately 1000 hours on that day, a 36-year-old professional pilot employed by a local air-taxi service entered the Fresno Flight Service Station. He did not talk to the duty forecaster but was observed standing at the self-help briefing counter. About an hour later, the pilot contacted the FSS by telephone and filed an IFR flight plan from Fresno to Elko, Nevada. The pilot stated that he had received the current weather information for the intended flight. His flight-plan route was Victor 23 to Linden VOR, Victor 113 to Reno, and Victor 6 to Elko. These airways lead over high mountains, across Lake Tahoe, and include a minimum altitude requirement of 13,000 feet along one segment of Victor 13 and a minimum terrain-clearance altitude of 12,000 feet.

At 1041, Cessna 206 N5085U called Fresno Tower for taxi instructions and IFR clearance. Aboard with the pilot were two passengers, one a 17-year-old student pilot who was going on the flight to gain practical experience. The aircraft departed at 1100 and climbed eastward toward a most formidable opponent—a mountain snowstorm. The flight proceeded normally under radar control along Victor 23, although several times the controller reported losing positive identification because of an intermittent transponder signal. At 1136, N5085U reported passing Woodward Intersection, 75 miles northeast of Fresno, at an altitude of 8,000 feet. The following is a transcript of the relevant radio contacts as recorded by Oakland Center (SKN = Oakland Center Radar, Stockton Sector; RNO = Oakland Center Radar, Reno Sector; 647 = United

Airlines Flight 647; 85U = Cessna 206 N5085U; * indicates communications between ground stations on direct land line).

TIME: 1136
85U—Oakland Center, Cessna 85 Uniform over Woodward Intersection.
SKN—Cessna 85 Uniform, roger, radar contact.
85U—Eight-five Uniform.
SKN—Cessna 85 Uniform, climb and maintain 12,000.
85U—We're out of 8,000 for 12,000.

TIME: 1145
SKN—Cessna 85 Uniform, Stockton altimeter 30.19, continue climb, maintain 14,000, over.
85U—One-four thousand, roger.

TIME: 1152
SKN—Cessna 85 Uniform, say altitude.
85U—Eight-five Uniform, passing 11,500.

TIME: 1153
SKN—Cessna 85 Uniform, Reno altimeter is 30.02.
85U—Three zero zero two, roger. We've reached 12,000—light icing and very light turbulence.
SKN—Roger.

TIME: 1201
85U—Oakland Center, 85 Uniform, request, ah, 12,000 for an altitude, ah, icing is, ah, starting to (unintelligible).
SKN—Eight-five Uniform, roger, we'll check and see if we can get you a lower.
85U—Roger, sir, I'm, ah, at, ah, according to altimeter setting, at 11,900 present time.
SKN—You say you're at 11,000?
85U—One-one nine.
SKN—Roger.
SKN—Cessna 85 Uniform, maintain 13,000, over.
85U—One-three thousand, roger.

TIME: 1204
SKN—Cessna 85 Uniform, radar contact lost, report Spring Hill.
85U—Roger, report Spring Hill.

TIME: 1206
SKN—Cessna 85 Uniform, request condition of icing now.
85U—Ah, icing, ah, negative at present time; we're between layers, ah, estimating, ah, along, ah, one-eighth of an inch of rime, over.
SKN—Roger, contact Oakland Center 128.8.
85U—One-two-eight point eight, thank you, sir.

TIME: 1210

85U—Oakland Center, Cessna 5085 Uniform, altitude, ah, 12,000 in the vicinity of Spring Hill.

RNO—Cessna 85 Uniform, roger, understand you are at 12,000, is that correct?

85U—That is affirmative, sir; I have, ah, picked up a little ice and, ah, very, very slow climb.

RNO—Cessna 85 Uniform, roger, ah, you realize you are the minimum obstruction-clearance altitude at 12,000?

85U—Understand at MOCA 12,000.

RNO—Roger, are you going to continue on, ah, your flight plan, or are you going to start going somewhere to get rid of the ice?

85U—Oh, I'll attempt the second leg of, ah, Victor 113; if negative, ah, climb, I will reverse course then to lose ice.

RNO—Roger.

TIME: 1213

RNO—Cessna 85 Uniform, report passing Richardson Intersection.

85U—Ah, roger, sir, I'm going to reverse course and get out of this condition. I want to reverse course and head back toward the Linden VOR on Victor 113.

RNO—Roger, advise passing, ah, Spring Hill on your way back.

85U—Eight-five.

TIME: 1216

*RNO—Stockton, Reno?

*SKN—Yeah.

*RNO—Cessna 85 Uniform is losing altitude, he's going to, ah, reverse course, and I told him to, ah, report Spring Hill to me . . . I don't have him on radar. He's now at 12,000, block 13 and 12.

*SKN—Okay, and I've got a possible target, ah, 15 northeast of Spring Hill.

*RNO—So do I, but he's got no DME, so I have no way of knowing if that's him. It looks like it might be the guy.

*SKN—Okay.

*RNO—Yeah, he's just reversed, that's the guy.

*SKN—Roger.

*RNO—All right, but I'm not going to take radar on him.

*SKN—Yeah.

*RNO—I'm not that positive. But he will report Spring Hill to me, and I'll turn him over to you.

*SKN—Right.

TIME: 1217

*RNO—I don't know where he's going or what he wants to do, and I don't want to bother talking to the guy . . .

*SKN—Okay.

*RNO— . . . while he's in these conditions.

TIME: 1218
RNO—Cessna 85 Uniform, ah, you can climb to 13,000 at any time you're able to; you can, ah, cruise at 13,000—13,000.
85U—Roger, sir.

TIME: 1220
RNO—Cessna 85 Uniform, how are you coming along?
RNO—Cessna 85 Uniform, how is the icing condition coming now? Cessna 5085 Uniform, Oakland Center, how do you hear?

TIME: 1223
RNO—Cessna 85 Uniform, Oakland Center.

TIME: 1225
*RNO—Stockton, Reno?
*SKN—Go ahead.
*RNO—We've lost radio communications with Cessna 5085 Uniform.
*SKN—Roger.
RNO—United 647, would you try to give Cessna 5085 Uniform a call?
647—Cessna 85 Uniform, Cessna 85 Uniform, how do you read? Center is trying to call you; we'll relay.
RNO—Six-Four-Seven, do you have contact with him?

TIME: 1228
647—Oakland Center from United 647, ah, we have him up and he said something about a 90 degree heading to the Linden VOR and I think he said something about icing. Stand by and I'll get what he has to say.
RNO—Okay, and United 647, squawk 1100 normal.
647—Ah, go ahead, 85 Uniform. Okay, roger, ah, center, ah, 85 Uniform says he is down to 7,500 feet and he can see the ground and he is on a 120 [sic] heading to the Linden VOR.
RNO—United 647, roger, thank you, and would you tell him to contact Oakland Center on, ah, 120.2?
647—One-two-zero point two.
RNO—That is correct.
647—Eight-Five Uniform, ah, want you to contact the Oakland Center on 120.2.
*RNO—Stockton.
*SKN—Radar.
*RNO—And, ah, 85 Uniform is down to 7,500 feet, he's got your frequency.
*SKN—Seven-five hundred?
*RNO—And United 647 is talking to him.
*SKN—Okay.
*RNO—And he's on a heading 210 for Linden.
*SKN—Not good.
647—I think he quit center, ah, he doesn't answer.

22

RNO—Okay, thank you.
*RNO—All right, that's his, he's—he's, ah, that's his heading, 210 to Linden.
*SKN—All right.

No further contact was made with the aircraft. Search and rescue units were immediately notified, but the severe storm made flight over the area impossible for several days. After the front passed, the higher mountains and canyons continued to be covered by low clouds that greatly limited the search. After more than a week of concentrated effort, the search was finally canceled, and the aircraft was considered to have crashed with fatal results for the occupants.

Although the official search had ended, a group of pilots from the Fresno area solicited contributions to continue the attempt to locate the aircraft. On January 26, a full 15 days after the crash, a helicopter hired for the private search was proceeding along the top of a ridge near Spring Hill Intersection when they spotted the figure of a man waving frantically to attract their attention. The helicopter landed and picked up the seriously injured 17-year-old passenger on N5085U. He directed the search aircraft to a point five miles northeast of Spring Hill, where they found the aircraft buried in deep snow among a stand of large trees. The wreckage was so positioned that the searchers were unable to see it while hovering directly overhead. In the cabin, they found the second passenger, who had suffered major injuries, huddled in the cold, unable to walk. The body of the dead pilot was still strapped in his seat. He had been killed on impact.

The youthful passenger described the moment of impact: "In the 180-degree left-turn, we lost a lot of altitude. I believe the last altitude I saw was 7,500 feet. We then saw the ridge and trees directly ahead. The plane was still level but losing altitude. I believe the airspeed was indicating about 80 mph, with full power. The pilot attempted a left-turning pull-up, but we hit the trees." The other pilot described their actions after the crash: ". . . we got out of the aircraft but decided to get back in because it was snowing hard and we were getting covered with snow. This particular snowstorm lasted about two days. We had about one-half day of good weather, then the low clouds covered us or were right down to our level. During this time search aircraft could be heard, but they were apparently flying the valley below us."

Investigation at the scene revealed that the aircraft was lying on its left side with both wings separated. The throttle was full open, propeller control in high rpm, and the flaps fully extended. All evidence indicated that the engine was developing power at impact.

Since the weather conditions that were present along the intended route of this flight were well-defined in all the available reports, it can only be assumed that the pilot did an incomplete job of self-help briefing and either overlooked or ignored the numerous and obvious warnings of the severe weather along his course. He certainly made a gross error in judgment by attempting to penetrate the active frontal system in an aircraft equipped only with a heated pitot tube.

In spite of this, the pilot had ample warning of the icing problem long before the problem became critical to the safety of flight. The most obvious of these opportunities came at 1210, when the pilot chose to continue on his course even though he was unable to climb to the assigned altitude of 13,000 feet and was struggling to maintain 12,000 feet—the minimum obstruction-clearance altitude for this airway. Three minutes later, the course reversal was finally made but too late to save the aircraft. At this point, the ice had accumulated to such a degree that 4,500 feet of altitude would soon have been lost. Now the Cessna had dropped into a mountainous area where the high ridges were dotted with peaks rising over 10,000 feet.

The fact that the two passengers spent more than two weeks exposed to subfreezing temperatures without food, heat, or adequate shelter and were still alive is nothing short of miraculous. The pilots who took it upon themselves to finance and continue the private search are certainly to be commended, and their efforts, no doubt, saved two lives. The final, desperate effort of the young passenger to walk out of the crash area also made the rescue possible. It is extremely doubtful that the crumpled airplane would have been sighted because of its hidden location.

At the time when the pilot of N5085U turned back to Spring Hill Intersection, his situation had deteriorated to the point where there is serious doubt that any action to change the final outcome was possible. Moments later center had information that the aircraft was at 7,500 feet and proceeding back to Linden VOR via Spring Hill Intersection. The ground level at Spring Hill is 7,621 feet making collision with the ridge inevitable if the pilot continued on his course and altitude. A turn to a heading of 250 degrees leads directly away from the highest terrain and parallels the ridges as the mountain slopes drop sharply into a broad opening valley. Realistically, the pilot himself should have been aware of the general terrain features and should have taken the proper action to extricate himself from the desperate situation. Instead he followed standard procedures and headings that led to an almost unavoidable crash.
—W. L. T.

Under the circumstances, the pilot's failure to make the 40-degree right turn that would have avoided the ridge at Spring Hill is not hard to understand: very few instrument pilots carry open VFR charts beneath their IFR charts for quick reference, and it's unlikely that in a rapidly deteriorating situation the pilot could have made much sense of the contour chart anyway. Most likely he was at, or barely at, 12,000 feet at 1210 when he passed Spring Hill and lost altitude steadily from before the point at which he decided to turn back until 1228, when he relayed an altitude of 7,500 feet near Spring Hill.

Ice is an insidious enemy. It attacks on two flanks, adding weight to the airplane and at the same time ruining its aerodynamic shape. As ice accumulates, more and more power is necessary to maintain speed and altitude, and the pilot gradually finds himself forced to sacrifice first his speed and then, bit by bit, his altitude in a desperate struggle to stay airborne. He confronts two approaching limits: the minimum, or "stalling," speed of the airplane, below which it cannot fly, and the altitude of the terrain below him. In order to operate safely in icing conditions, airplanes must be equipped with devices to keep ice from accumulating or to slough it off once it has accumulated, and they must be sufficiently powerful to carry a load of ice or to climb rapidly through the levels at which icing is occurring to drier, colder altitudes, often above 15,000 feet, where ice no longer forms. The Cessna 206 had no anti-icing equipment to speak of and no great reserve of power; the pilot who does battle with ice in such an airplane is depending on luck alone.

The more experienced a pilot is, the more aware he is of the uncertainties of the prediction of icing conditions and of the possibility of flying without any icing in an area where an hour before heavy icing was reported. Few professional pilots actually adhere to hard-and-fast rules of when to go and when not to go. When conditions are uncertain, they rely as often as not on intuition. In this case, the conditions were not so much uncertain as positively awful, but the pilot probably allowed the prospect of improving weather beyond the Sierra Nevada to lure him into taking a chance of squeezing over the top. Another pilot, in the following incident, also highly experienced and flying in the same area was waylaid not by icing but by a mechanical failure. He reacted at first with perfect calm.—P. G.

3.

"UH, HAVING A LITTLE TROUBLE"

A Cessna 310 N5811X departed McCarran Airport, Las Vegas, Nevada at 0950 on the morning of February 19, 1969 on an IFR flight plan to Reno, Nevada. Aboard the aircraft with the pilot were his wife and a married couple who had accompanied them on a vacation from their home in Seattle, Washington. The flight was the first leg of the return trip. After departure the Cessna proceeded under normal IFR routing. At 1111 the aircraft entered the control area of Oakland Center. The following is a transcript of the pertinent radio contacts concerning N5811X (11X = Cessna N5811X; OAK = Oakland Center; RNO = Reno Approach Control; * denotes communication between ground stations).

TIME: 1111
11X—Oakland Center, this is 5811 Xray, 127.3, over.
OAK—Five-Eight-One-One Xray, Oakland Center, go ahead.
11X—One-One Xray, we would like to make a position report.
OAK—Go ahead.
11X—One-One Xray. Coaldale at zero nine, 13,000, IFR Reno, five six Reno.
OAK—One-One Xray, Oakland Center, roger. Squawk 1100.
11X—One-One Xray, again, please.

TIME: 1117
OAK—Five-Eight-One-One Xray, ident.
11X—Identing.
OAK—One-One Xray, radar contact.

TIME: 1133
OAK—One-One Xray, Oakland Center, what is your flight condition in your present position?
11X—The weather is, uh, the flight conditions are fine. We're at 13,000, it's smooth, I'm IFR.
OAK—One-One Xray, roger.
11X—One-One Xray, there is no icing.
OAK—Roger, thank you. Any appreciable turbulence?
11X—No, it's smooth, occasionally just a very light ripple, that is all.
OAK—Okay, thank you much.
11X—Oakland Center, this is twin Cessna 5811 Xray, we would like to

26

continue our VF, uh, excuse me, our IFR flight plan to Lakeview, Oregon, Victor 165. And we estimate, uh, our flight plan as one hour and five minutes over Reno, over.

OAK—One-One Xray, roger, expect a clearance to Lakeview shortly.

11X—Lakeview, Oregon.

OAK—Cessna 5811 Xray, Oakland, clearance to the Lakeview Airport via Victor 105 Reno, Victor 165 Lakeview, and climb and maintain 14,000, over.

TIME: 1137

11X—One-One Xray is cleared to, uh, Lakeview, Victor 105 and Reno Victor 165 Lakeview, start climb and maintain 14,000. We are out of 13,000 at this time, over.

OAK—One-One Xray, roger.

TIME: 1140

11X—Oakland Center, twin Cessna 5811 Xray is now level 14,000.

OAK—Twin Cessna 5811 Xray, roger, 14,000.

TIME: 1143

OAK—Twin Cessna 11 Xray, contact center on 128.8.

11X—Oakland Center, twin Cessna 5811 Xray, 128.8, over.

OAK—Cessna 5811 Xray, roger, loud and clear.

11X—One-One Xray.

OAK—Cessna 11 Xray, the Reno altimeter, 29.62.

11X—Two niner six two, and would like, rather than landing at Lakeview, to land at Klamath Falls. We can get out of Victor 126. If you could vector us over to Klamath, we would appreciate it.

OAK—One-One Xray, stand by, let me check. Cessna 11 Xray, can you accept 16,000?

TIME: 1200

11X—Center, we would rather stay at 14,000; we think we have lost our, uh, having a little trouble with our number-two engine at this time. Let me just check here a minute.

OAK—Okay, sir, 5811 Xray, Oakland Center, you're cleared to the Klamath Falls Airport via Victor 165 and then via radar vector by the Seattle Center to, uh, Klamath. Maintain 14,000, go ahead.

TIME: 1201

OAK—Cessna 11 Xray, say your altitude.

11X—One-One Xray is 11,600.

OAK—Cessna 11 Xray, contact Oakland Center on 134.65.

11X—One three four point six nine?

OAK—Point six five.

11X—One-One Xray, understand maintain 14,000 on Victor 165, then radar vector from Seattle Center to Klamath Falls, thank you.

OAK—One-One Xray, that is correct. In the event of lost communications, proceed via Victor 165 Lakeview and your clearance limit will be Lakeview, go ahead.

11X—One-One Xray, in the event of lost communications, proceed to Lakeview.

OAK—Affirmative.

TIME: 1219

11X—Oakland Center, Cessna 5811 Xray, we would like to feather—our right engine is out and we feathered. We're losing some altitude; we have a blower on the left engine. I think we will go ahead and continue, uh, we're 25 miles DME out, over.

OAK—One-One Xray, affirmative, I have you still in radar contact four miles north-northwest of Bonham Intersection. One four thousand is the MEA, sir, and, if you go much lower than that, I may lose radar contact with you.

11X—One-One Xray, I think we'll go ahead and try to go on from here, over.

OAK—Roger, understand, are you going to cancel out?

TIME: 1220

11X—No, no, we are anxious to try it ahead, and try to keep with us.

OAK—One-One Xray, roger, uh, I am beginning to lose you on radar now, uh, because I think you may be descending below 14,000 slightly. I would suggest you return to Reno, sir.

11X—Okay, we'll return. We are doing, uh, the 180 toward the (unintelligible).

OAK—One-One Xray, I still have you radar contact. You are cleared to the Reno Airport via Victor 165 and maintain, uh, 14,000. I can get you 12,000 in about five miles as soon as you get back a little bit southeast.

11X—One-One Xray.

*OAK—And Reno, you got pretty good weather now?

*RNO—Our weather is sky partially obscured, 900 scattered, measured ceiling 1,200 overcast. Visibility at three. Light rain showers, rain or snow showers rather than fog; however, an aircraft just reported here a few minutes ago, northbound, then northeast, said he'd seen the ground at 9,000.

*OAK—Okay, I'll give them the weather.

TIME: 1222

OAK—Twin Cessna 5811 Xray, Reno Approach Control advises that the weather in Reno, 900 scattered, 1,200 overcast, with light rain showers and some fog; however, he had an aircraft just report that he could see the ground from 9,000 feet.

11X—And we're at 10,500 at this time.

OAK—One-One Xray, can you maintain 10,500, over?

11X—We'll do the best we can, over.

OAK—Roger, sir, I am unable to assign an altitude below 11,000 in your present area, so maintain the best you can.

11X—Okay, 11,000, 11 Xray.

*OAK—Reno Approach, here is an attempted radar hand-off on 5811 Xray. The one who just lost an engine is over Bonham at 10,500, and he is maintaining

the best altitude he can and squawking 1100.
*RNO—Don't have anything. Just clear him for a straight-in ILS 16 approach. Pyramid Intersection direct.
*OAK—Okay.

TIME: 1224
OAK—One-One Xray, Oakland Center, clearance for straight-in ILS, clearance to the Reno Airport via Pyramid, direct Reno, go ahead.
11X—One-One Xray, cleared for ILS from Pyramid direct Reno.
OAK—One-One Xray, affirmative, Reno Approach Radar does not have you in radar yet; I am attempting hand-off.
11X—One-One Xray, we're at 9,000, over.
*OAK—Roger, Reno Approach, Oakland, 11 Xray is down to 9,000—doing the best he can.
*RNO—Okay, what type of Cessna is it?
*OAK—It's a Cessna 310.
*RNO—Okay.
*OAK—And let me know as soon as you see something up there.
*RNO—Okay.

TIME: 1225
OAK—Twin Cessna 11 Xray, Oakland Center, say your altitude now, please.
11X—I'm at 8,000.
OAK—Roger, can you maintain anything higher than eight?
11X—Negative, negative. We've just hit our stall warner.
OAK—Okay, sir.

TIME: 1226
*OAK—Reno Approach?
*RNO—Reno.
*OAK—Okay, this 11 Xray is down to 8,000 now, and I estimate him to be somewhere between Bonham and Pyramid. I have lost him now on radar.
*RNO—We don't see anything up there either.
OAK—Okay, 11 Xray, Oakland Center, radar lost, say your heading.
11X—One-One Xray, heading 120.
OAK—One-One Xray, roger, you may be approximately 10 to 12 miles north of Stead Airport. Keep your eyes out for it, sir. You may be able to get in there.
11X—We'll look.

TIME: 1227
OAK—Okay, Stead should be slightly off, uh, to your right, approximately one o'clock position, oh, eight or 10 miles.
11X—On our right, you say?
OAK—It should be approximately on your right.
11X—(Unintelligible transmission).
OAK—One-One Xray, Oakland Center, uh, squawk emergency, squawk 7700.

TIME: 1228
*OAK—Reno Approach, Oakland, do you see anything of 11 Xray as yet?
*RNO—Negative.
*OAK—Okay.
OAK—One-One Xray, Oakland Center, radio check. Twin Cessna 5811 Xray, Oakland Center, radio check, how do you hear?

When radio contact with the Cessna was lost, an immediate search was initiated for the missing aircraft. At noon on the following day, the wreckage was located on the steep western slope of the 8,722 foot Virginia Mountain. The aircraft had struck the slope at an altitude of 5,900 feet. The left wing-tip fuel tank, which was crushed to about half its length, was the first piece of wreckage found along the impact scatter path. Parts and pieces of the aircraft were located for a distance of 55 feet in a direction of 315 degrees magnetic, downhill from the main wreckage, which was facing in a direction of 025 degrees magnetic. A severe postimpact fire had consumed much of the cabin-cockpit area, making an accurate determination of the instrument readings or flight- and engine-control settings impossible. The fire had also consumed all baggage, making a computation of the weight unobtainable. However, it was found that the baggage included several suitcases, some small overnight cases, and three sets of golf clubs.

Examination of the powerplants revealed that the left engine was operating at the time of impact; the right-engine propeller was feathered. Cause of the failure of the right engine was found to be internal structural damage. A large hole was evident in the top left crankcase half adjacent to the number-six cylinder. A smaller hole was found in the right crankcase half near the number-five cylinder. Further investigation showed the absence of the number-five piston and rod. The remains of these parts were located in the oil sump. The rod and cap displayed evidence of severe pounding, indicating that power had remained on for some time after the initial failure.

Although the pilot of N5811X encountered an emergency under difficult circumstances—the loss of an engine during IFR flight—under normal conditions he should have had no serious problems in holding a safe altitude for the return to Reno. Unfortunately, this was not the case. Statements made by the pilot during the emergency make it clear that he was unable to maintain even level flight at an altitude well below minimum terrain clearance. Factors contributing to the inability of the Cessna to hold its altitude during single-engine operation may be one of several or a combination of many factors.

First to be considered is excessive gross weight, either in payload or airframe ice accumulation. Since the pilot reported no problem with icing at any point in the flight, it is unlikely that the latter condition was present. Unfortunately, the fire prevented any accurate calculation of the baggage weight, and the performance of the aircraft during the more than two hours of flight before the accident, plus the considerable weight reduction due to fuel burn-off, makes an overweight or out-of-balance situation highly unlikely.

Pilot technique during single-engine procedures, particularly during actual IFR flight, can have a great influence on aircraft performance. The pilot's experience level—2,894 hours, 514 of which were in Cessna 11X, plus 311 hours of actual IFR—would indicate that this commercial pilot, with single-engine and multiengine land and instrument ratings, possessed adequate training and experience to cope with the problems.

The Cessna's engines were equipped with a popular turbocharger conversion, modification accomplished in compliance with an STC approved by the FAA. According to an engineer employed by the company providing this installation, the aircraft should have been capable of maintaining a single-engine altitude of 16,000 to 18,000 feet at its gross-weight configuration. It should be noted that performance charts concerning this modification are not available, and none is required by the FARs. The fact that the manufacturer's performance claims were not verified by actual test flight is also noteworthy. However, with or without the engine modification Cessna 5811X should have had the capability to return easily to Reno on one engine. Since this was evidently not possible, the pilot must have faced another problem, perhaps one that he was not even aware existed.

The initial cruising altitude assigned to 5811X upon departure from Las Vegas was 13,000 feet. After crossing the Reno VOR, the flight maintained 14,000 until the engine was lost. During this time, the pilot was flying at high altitudes, evidently without the use of oxygen. No supply of oxygen was found in the wreckage, although the aircraft was partially equipped for such a system. It is possible that the pilot was suffering some effects of hypoxia that were to influence his reactions to the emergency about to occur.

Whatever the cause, and despite the fact that 11X could not hold an adequate altitude, the accident need not have happened. When the flight crossed the Reno VOR on course at 1156, its altitude was 14,000 feet. Exactly four minutes later the pilot stated that he was experiencing trouble with the right engine. One minute later he reported his altitude as 11,600 feet, an abrupt change of 2,400 feet. From the evidence later found there is little doubt that partial failure of the number-five piston assembly had occured at this point, yet the pilot continued on course into higher terrain.

Finally, at 1220, with the failed engine feathered and the aircraft drifting below the required MEA, the pilot still opted to continue toward his destination more than 150 miles ahead. Only after a suggestion by the controller did 11X reverse course for the return to Reno, now some 45 miles behind. At 1222, with radar contact lost, the pilot reported that he was established on the reverse course at an altitude of 10,500 feet. Three minutes later, the pilot gave his altitude as 8,000 feet—still descending. By now, the aircraft had drifted below the numerous peaks scattered along the return course to Reno. At 1227, the controller suggested to the harried pilot the possibility of spotting Stead Airport, located about 10 miles north of Reno. This may have influenced the pilot to descend farther. The last transmission from 11X came just after 1227. The pilot's voice was high-pitched, garbled and unreadable, possibly transmitted the instant before impact. When the left wing of the Cessna struck the mountain, its heading was approximately 315 degrees, a course leading away from Reno.

What would influence the desperate pilot to take a heading away from his only safe airport? Was he descending those last few minutes in hope of finding Stead Airport? Or had oxygen starvation robbed his mind of the ability to determine and to hold an accurate and constant heading? The facts needed to answer these questions will probably never be known. Perhaps the most significant lesson to be learned from this tragedy is the necessity of taking immediate action at the first indication of actual or suspected trouble. First be certain that you have the ability to evaluate and judge the situation. Hypoxia is a stealthy and ever-present threat at higher altitudes. Just because you haven't yet met it at 10,000 or more doesn't mean that you won't the next time. The best time to prepare for trouble is before takeoff. Minor problems have an uncomfortable habit of snowballing into major problems, and major problems have a terrifying tendency to grow into uncontrollable monsters.—W. L. T.

Why didn't he turn back to Reno and land immediately? Was it in order to allay the fears of his passengers that he continued as though nothing had happened? Since the airplane was capable of sustained flight at high altitudes on one engine, one is indeed tempted to wonder whether the pilot intentionally descended below 9,000 feet, lured by the report, made some time before, of another pilot's having seen the ground from 9,000 feet. Again, the pilot of 11X probably did not have VFR contour charts before him; perhaps he took a chance on the terrain, encouraged by the fact that the few terrain heights given on the IFR charts—the elevations of airports—are comparatively low, all of them below 5,000 feet. The failure of the airplane to maintain height in the first place is still a mystery. Perhaps downdrafts carried it down, or perhaps the pilot did not realize at how low an airspeed he would have to fly in order to maintain altitude. There is still a piece missing in the puzzle—something to account for the drastic change in the pilot's state of mind between the time when he calmly announced that an engine had quit and 27 minutes later when he flew, apparently badly disoriented, into a mountain. Hypoxia—the insidious oxygen starvation that is nearly impossible to detect in oneself—may be the missing link; perhaps it was the emotional atmosphere in the cockpit—a terrified passenger, an altercation—impossible to know.—P. G.

4.

"UNABLE TO MAINTAIN . . ."

Even though the pilot of the pressurized twin elected not to fill the fuel tanks, the aircraft was probably 225 pounds over gross when it departed Eagle, Colorado on a Friday afternoon with six and baggage on board. The departure for Las Vegas, 448 nm away, got off to a faltering start, as one engine balked and had to be heated before it would start. Next, the baggage door came open on takeoff, and the pilot had to return, repack the nose compartment, add a bit more fuel, and then take off again. The flight was finally off at 1703 MDT, IFR, filed altitude 14,000; estimated time en route to Las Vegas two hours and 20 minutes; fuel on board three hours and 10 minutes.

One hour and seven minutes after takeoff, the pilot notified the controller that he would be requesting a lower altitude, 12,000 feet, after passing Bryce Canyon. The pilot stated that winds aloft were worse than forecast and that fuel reserves were a little low for legal reserves to Las Vegas. The pilot subsequently had some conversation with another aircraft on the frequency about ground-speed readouts, reported that his DME was intermittent, told of a 48-knot headwind component at his altitude, and requested en-route weather. The aircraft, which we will refer to as 35D, contacted Los Angeles Center at 1835 MDT per instructions. The following transcript covers the pertinent conversation during the next 38 minutes of flight (LAX = Los Angeles Center; N66R = another aircraft on the frequency that relayed transmissions when communications between 35D and Center were not possible; all times are Mountain Daylight, based on the 24-hour clock).

TIME: 1835
LAX—November 35 Delta, Los Angeles Center: if you read, ident.
35D—Roger, I read.
LAX—Three-Five Delta, roger. Loud and clear. Verify 14,000, your Cedar City altimeter 29.80.
35D—Altimeter and, er, about 30 knots higher-than-forecast headwinds.
LAX—Three-Five Delta: I'm sorry, sir, you're still intermittent; uh, give me another call in a few more miles, please.
35D—Weather, over.
LAX—Three-Five Delta, Los Angeles. I'm sorry, yours, er, been broken out, er, still might be a little bit far. Say again, please.
LAX—Six-Six Romeo, are you reading 35 Delta okay?
66R—Affirmative.
LAX—Ask him what his message was and relay to me, would you please? He's a little bit too far out from our transmitter site.

33

66R—Three-Five Delta, say your message again.

66R—Los Angeles Center, 66 (unintelligible).

LAX—Six-Six Romeo, Los Angeles.

66R—Thirty-Five Delta informs that they're running into about 40 knots, uh, higher headwinds than anticipated.

LAX—Six-Six Romeo, roger. Advise him I have him in radar contact 45 miles northeast of Bryce Canyon.

66R—Three-Five Delta. Center advises they have you in radar contact about 45, er, DME north-northeast Bryce Canyon.

TIME: 1842

35D—Los Angeles Center, 35 Delta.

LAX—Three-Five Delta, Los Angeles.

35D—Three-Five Delta. We'd like a lower altitude at Bryce Canyon, one-two thou.

LAX—Three-Five Delta, understand your request. Squawk 1131, code 1131.

35D—Three-Five Delta, we have an inop DME. Would you advise our position north, uh, east of, uh, Bryce Canyon?

LAX—Three-Five Delta, I show you 30 miles northeast of Bryce Canyon.

TIME: 1849

35D—Los Angeles Center, 35 Delta.

LAX—Three-Five Delta, Los Angeles.

35D—Er, Three-Five Delta. Can we get that lower altitude now? We're picking up rime ice.

LAX—Three-Five Delta, er, maintain, er, 13,300, that's my MEA on the airway until you get to Bryce.

35D—Er, Three-Five Delta. Would you get us, uh, could you get us something, uh, on St. George weather, over?

LAX—I don't have the weather for St. George. Stand by. We'll contact, uh, Cedar City Radio.

LAX—Three-Five Delta, they advise, er, St. George weather is not available.

35D—Er, LA Center, er, 35 Delta. We're picking up ice, low on fuel, and, ah (unintelligible), ah, requested, ah, someone, ah, get hold of St. George (unintelligible) if you can get their weather, over.

LAX—Three-Five Delta, roger, ah, I'll try to go through, ah, flight service station.

TIME: 1852

35D—Three-Five Delta. Like a higher altitude at 14,000 if I can't get a lower one.

LAX—Three-Five Delta, you start climb now, and, uh, how high would you like to go, sir?

35D—You got a report on tops in this area?

LAX—Six-Six Romeo, do you have any, er, reports on the tops northeast of, uh, Bryce Canyon by any chance?

66R—Ah, negative. Er, once he gets beyond Bryce City, ah (unintelligible), 15,000 (unintelligible), 16,000 at this time.

LAX—Three-Five Delta, we don't have any reports on the tops. Ah, Six-Six Romeo was northeast bound at 15; he picked up icing also, and he said he broke out just southwest of Bryce Canyon. [Six-Six Romeo had discontinued northeast-bound flight and headed southwest due to icing at the time of this exchange.]

66R—And the tops southwest of Bryce Canyon are running right at 15,000.

LAX—Three-Five Delta, Los Angeles Center: you were blocked out, would you say again, please?

35D—Uh, Three-Five Delta. Do you have any (unintelligible); getting real heavy ice.

LAX—Six-Six Romeo, can you relay, uh, 35 Delta's request to me? I'm not picking him up at this time.

TIME: 1854

66R—Uh, Three-Five Delta, Los Angeles advises that, uh, he's unable to pick you up at this time; say your (unintelligible).

35D—Three-Five Delta. We're picking up heavy ice; we're trying to get above it.

66R—Roger, Los Angeles, uh, 35 Delta advises (unintelligible) ice (unintelligible), trying to climb above it.

LAX—Six-Six Romeo, thank you.

35D—Uh, Six-Six Romeo, do you have a position for, uh, 35 Delta?

LAX—Roger, 35 Delta's, uh, 13 miles northeast of Bryce Canyon at present time.

66R—Okay, 35 Delta, you're, uh, 13 miles northeast Bryce Canyon, and, uh, when you get to Bryce Canyon, you should be able to proceed in the clear, 15,000.

TIME: 1855

66R—Uh, Three-Five Delta, do you read me?

66R—Uh, Three-Five Delta, do you read 66 Romeo?

35D—Affirmative, I read you. I'm unable to get above one-four-five, 14,500.

66R—Okay, understand you're unable to climb above fourteen-five (unintelligible), Los Angeles.

LAX—Six-Six Romeo, understand he's at 14,500. Is that correct?

66R—He advises he's unable to climb higher at this time.

TIME: 1857

35D—Try a lower altitude. Delta, we'd like to try a lower than one-four; we're unable to maintain one-four.

LAX—Three-Five Delta, understand, you, ah, unable to maintain 14,000.

35D—Affirmative. Give our position northeast of Bryce Canyon, please.

LAX—I show you six miles northeast of Bryce Canyon; uh, descend and maintain 12,000; fly heading 220.

35D—Three-Five Delta, rog.

TIME: 1901

LAX—Three-Five Delta, say your altitude now.

LAX—November 35 Delta, say altitude.

35D—We're at 12,000.

LAX—Three-Five Delta, roger.

35D—Can you advise me Bryce, uh, Bryce Canyon weather, please?

LAX—November 35 Delta, Bryce Canyon weather is indefinite, 1,000, uh, obscured, one-half-mile visibility with light snow and fog.

35D—Three-Five Delta, roger. Would you advise us (unintelligible) arrive over Bryce Canyon now, over.

LAX—Three-Five Delta, I concur, I show you over Bryce Canyon at this time.

TIME: 1904

LAX—November 35 Delta, the Cedar City weather is 8,000, uh, correction, 800 scattered, estimated 1,500 broken, 3,000 overcast, visibility 10 miles, and we're getting the St. George weather through now.

35D—Roger, might make it to St. George. We're holding 12,000, uh, making, uh, about 130 knots, have a lot of ice, uh, I don't know if we can maintain this altitude, over.

LAX—Three-Five Delta, roger. And here's the St. George weather. It's estimated 1,000 overcast; nine miles, uh, visibility. It appears, uh, Cedar City a little bit better.

TIME: 1906

LAX—Three-Five Delta, uh, Cedar City Radio advises their weather good; remain as stated or improved.

LAX—Three-Five Delta, what are your intentions?

35D—Roger, just trying to, uh, hold my altitude.

LAX—Three-Five Delta, can you resume normal, uh, on Victor Eight?

35D—Negative, negative.

TIME: 1908

LAX—Three-Five Delta, request your fuel status.

35D—About one hour.

LAX—November 35 Delta, Los Angeles.

35D—Three-Five Delta.

LAX—Roger, sir, show you proceeding toward Cedar City. Your intentions?

35D—Ah, what's the weather at Cedar City?

LAX—Cedar City weather's 800 scattered, estimated 15,000, 1,500. It's estimated 1,500 broken, 3,000 overcast. Visibility is 10 miles.

35D—Do they have ILS there?

LAX—Three-Five Delta, negative.

TIME: 1911

LAX—Three-Five Delta, did you receive the message on ILS at Cedar City?

35D—Affirmative.

TIME: 1912

35D—Can you give us a radar steer to Cedar?

LAX—You have to climb to 13,000. I can give you a vector to Cedar, if that is possible.

35D—Ah, negative, we can't make 13,000.

TIME: 1913

LAX—Ah, Three-Five Delta, suggest heading, ah, 265, and, if possible, maintain 13,000.

35D—We're unable to maintain 12,000, maintain 13,000. We have, ah, too much prop ice to get up there.

That was the last recorded transmission from 35D. The other aircraft on the frequency reported transmission to the effect that 35D was down to 8,800-feet and that the pilot was just going to try to keep it flying. Contact was then lost. The aircraft was located four days later on the upslope of a mountain at the 8,100-foot level. All six occupants had perished in the crash and ensuing fire. The ELT aboard the aircraft did not function.

The pilot filed two hours and 20 minutes for the 448 nm trip, with three hours and 10 minutes' fuel on board. The time en route anticipated a groundspeed of 207 knots, which just about corresponds to the aircraft's level-flight, 75%-power cruise speed at 14,000 feet, with no allowance for climb, maneuvering, or headwind. The winds-aloft forecast called for light southwesterly winds at 12,000 but for much stronger winds at 18,000. The forecast headwinds were not reflected in the filed time en route. Presumably the pilot elected not to fill the fuel tanks because he knew that he already had an overload condition and didn't want to aggravate it further. Once off and en route, the pilot found the winds were stronger than he had planned for and that his groundspeed was approximately 160 knots. Allowing for some time to climb and maneuver, that should have told him that the flight to Las Vegas and the fuel on board would be close to equal. No early mention was made of a stop for fuel or of turning around.

At 14,000-feet 35D flew into the top of some clouds, probably stratocumulus, and into an area of precipitation. This is a situation most conducive to ice formation. The relationship between the airplane and the ice is somewhat of a mystery, because a nonturbocharged light twin flew through the same area at 14,000-feet not far behind 35D without any particular trouble. The pilot did report using airborne radar to avoid precipitation areas and surface and propeller deicing equipment to remove ice that formed; 35D was also equipped with radar and deicing. The pilot of 35D did not report any problem with his equipment, but it apparently did not function, or 35D flew through much more severe icing conditions than the other aircraft. The last is possible, because no mention was made of radar use or of deviations around areas of precipitation. The investigation showed nothing that would have rendered the prop or surface deicing inoperative, but, even so, severe conditions can create more ice than an airplane will carry. When 35D encountered icing, the aircraft had an hour and 25 minutes' fuel remaining and was about 199 nm from Las Vegas. At 160 knots it would have taken an hour and 15 minutes to reach Vegas without ice; with it, who knows?

It is important to evaluate the situation at the time of the first mention of ice. The pilot had an airplane that was capable of climbing 1,400 fpm at 14,000-feet, that had a service ceiling of 30,800-feet and that could have flown for almost two hours at reduced power at a higher altitude. If he had elected to reverse course at that time and to climb into the strong tailwinds, it would have been possible for him to throttle back for best economy and to ride with the tailwind back to Grand Junction, where there is an ILS. The landing there could have been made with almost an hour's fuel in reserve.

The only practical en-route stop ahead was Cedar City, where the ceiling was 1,500 feet, and the visibility 10 miles. That is quite good enough for the VOR approach, which has a minimum descent altitude of 344 feet agl. Bryce Canyon was closer, but the weather was below minimums. St. George was a possibility, but the ceiling there was 1,000 feet, and the MDA is 1,661 feet agl. In other words, when ice built to the point that the airplane's climb capability was seriously diminished, the pilot's options instantly narrowed to one—Cedar City. The trouble was that the country between the airplane and Cedar City was high and rough, and the airplane couldn't manage both the necessary altitude and the ice.—W. L. T.

Ice can form with incredible swiftness; in this case it apparently did. It is an adage among pilots that you should never be without an avenue of escape—never reduce your options to one. The pilot of 35D put himself into a precarious situation to start with; then, by delaying his decision to turn back, he checkmated himself; there was nothing to do but wait for the end.

Actually he had one remote possibility of salvation. If he had had a VFR contour chart available—this refrain recurs again and again—he might have turned toward the southeast, where lower, flatter terrain offered some chance of breaking out below the clouds before striking the ground. No controller could have been expected to suggest this course—it was tantamount to admitting that the pilot had only a 50-50 chance of being alive in a few minutes, and, besides, it is impossible for a controller to know the actual flight conditions or the capabilities of the airplane or the state of mind of the pilot.

It is easy in retrospect to say that the pilot delayed too long until the situation was desperate, but, when one has flown for a long time and experienced the infinite variety of chance and weather, one finds it easier to understand his hesitation. Unwilling to admit that he had planned the flight with inadequate fuel margins, encouraged by the successful passage of other planes through the area, and surprised by the sudden buildup of ice, he wavered: the icing might have stopped as rapidly as it had started, and turning back would have meant flying back through an area of known heavy icing. The required change of mind can be incredibly difficult. A basic rule says that, when in doubt, you should turn back, but it is precisely in order to escape such burdensome necessities that big, powerful, pressurized, turbocharged twins with deicing equipment are built and bought. Most of the time the equipment is up to the job; sometimes it isn't, and then the pilot's problems are twice as severe.

In the following account, another pilot, another airplane, another part of the country—but that same enemy, ice: if the pilot of 35D was too slow in deciding to back out and get onto the ground, the pilot of Aero Commander 61B was perhaps too hasty.
—P. G.

5.

THE FATAL DESCENT

For several days before making a long-planned vacation flight to Florida and the Bahamas, the pilot of an Aero Commander 560A, N2761B, had spent considerable time at the Weather Bureau self-briefing counter at Ross Field, Benton Harbor, Michigan. At 0700 on the morning of March 29, 1969, the pilot called the South Bend, Indiana FSS and requested a weather briefing for an IFR flight from Benton Harbor to Atlanta, Georgia. He was advised that his intended route would require him to penetrate an active cold front lying directly across his course. Considerable rime icing, turbulence, and areas of heavy precipitation were reported to lie within this front. The pilot was also given Airmets Bravo Seven and Charlie Six, which warned of low ceilings and visibilities developing in the mountains of east and central Tennessee.

The pilot arrived at the airport at 0930 and requested fueling of his aircraft; he asked that a special effort be made to see that the tanks were completely full. Fueling records show that 84.6 gallons of gasoline were required to bring all tanks to the maximum total capacity of 156 gallons. The Commander left its home base at 0950, IFR to Atlanta. On board with the pilot were his wife, two young sons, and father-in-law. After a routine climb clearance with radar vectors to the flight-plan course, departure control handed the flight off to Indianapolis Center. The following is a transcript of radio contacts with N2761B (61B = Aero Commander N2761B; IND = Indianapolis Center; 07Y = Airlift N907Y; ATL = Atlanta Center; UN = Unknown aircraft).

TIME: 1026
61B—Indianapolis Center, Aero Commander 2761 Bravo, climbing to maintain 9,000, over.
IND—Aero Commander 2761 Bravo, roger and, ah, squawk ident.
61B—Six-One Bravo, ident.
IND—Airlift 401 and 2761 Bravo all in radar contact.
61B—Six-One Bravo.

TIME: 1045
07Y—You might make a note that we've got light rime icing up here in the clouds.
IND—Zero-Seven Yankee, roger.
61B—Ah, 61 Bravo, ah, we're also picking up light ice now.

TIME: 1052

61B—Ah, 61 Bravo, ah, picking up light rime ice; what have I got ahead, over?

IND—Ah, 61 Bravo, I've had reports that at 10,000 you break above the weather.

61B—Ah, you say 10,000 feet is above the weather?

IND—Ah, this is what I had prior to this here; they say that 10,000 we're breaking out of the freezing level and, ah, 11,000 they were on top.

61B—Ah, would you give me 11,000 now?

IND—Two-Seven-Six-One Bravo, climb and maintain 11,000.

TIME: 1053

61B—Leaving niner for 11,000 now, thank you.

TIME: 1059

IND—Ah, 2761 Bravo, are you at 11,000 now?

61B—Seven-Six-One Bravo, level at 11,000.

IND—Roger, and what's your flight conditions at 11?

61B—Ah, like you say, we're on top by about 500 feet, ah, and, ah, we got, there's an overcast above us, ah, no rain.

IND—Two-Seven-Six-One Bravo, roger.

TIME: 1107

IND—Aero Commander 2761 Bravo, contact Atlanta frequency 127.55 now.

61B—Six-One Bravo, say the frequency again.

IND—Roger, Atlanta Center 127.55 for 61 Bravo.

61B—Six-One Bravo, 127.55 and thank you.

IND—Roger, good day, sir.

TIME: 1108

61B—Atlanta Center, this is Aero Commander 2761 Bravo, level at 11,000, over.

ATL—Aero Commander 2761 Bravo, Atlanta Center, roger, 11,000, squawk ident.

61B—Six-One Bravo.

ATL—Aero Commander 2761 Bravo, Atlanta Center, radar contact, you're cleared to the Dallas Intersection via Victor 51 West Chattanooga, then via the Chattanooga 180 radial and the 324 radial to Dallas, maintain 11,000.

61B—Uh, 61 Bravo, clear—cleared to Dallas Intersection via Victor 51, and I'll have to have the rest of that clearance, over.

ATL—Yes, sir, present routing to Chattanooga Victor 51 West to Chattanooga, then fly the 180 radial of Chattanooga to intercepting the Atlanta 324 radial to Dallas.

TIME: 1109

61B—Six-One Bravo, stand by.

ATL—Six-One Bravo, roger, uh, you are landing Atlanta, are you not, sir?

61B—I'm landing in Atlanta, roger.

ATL—Uh, roger, this is to, uh, get you routed in a new routing to Atlanta.

TIME: 1117

61B—Atlanta Center, this is Aero Commander 61 Bravo, over.

ATL—Aero Commander 61 Bravo, Atlanta Center, go ahead.

61B—Uh, 61, uh, Bravo, uh, I can't seem to locate your Dallas Intersection, over. Uh, can you give me some information on it?

ATL—Yes, sir, the Dallas Intersection is made up of the Atlanta Vortac 324 radial and the Norcross 264 radial, that is the Norcross 264 and the Dallas—and the Atlanta 324, over.

61B—Six-One Bravo, would you give me my clearance, and, uh, read it slow, please?

TIME: 1118

ATL—November 61 Bravo, roger, after Chattanooga, fly heading 180, heading 180, to intercept the Atlanta 324 radial; fly the Atlanta 324 radial to Dallas Intersection, and Dallas is the 29, 29-mile DME fix, over.

61B—Aero Commander, uh, cleared to fly the Chattanooga radial, 180 degrees to Atlanta 324-degree radial, 324-degree radial to, uh, Dallas Intersection, over.

ATL—Six-one Bravo, that is correct, that is correct.

61B—Six-One Bravo, thank you.

61B—Uh, 61 Bravo, uh, we're picking up rime ice, uh, pretty bad here, uh, is there any possibility of getting down to Crossville; what is the weather at Crossville, over?

TIME: 1119

ATL—Six-One Bravo, roger, descend and maintain 10,000 now and, uh, understand you'd like to land Crossville?

61B—Uh, 61 Bravo, I'd like to land at Crossville if, uh, conditions, uh, permit, over.

ATL—Six-One Bravo, roger.

61B—Six-One Bravo, leaving 11 for 10.

TIME: 1120

ATL—November 61 Bravo, Atlanta, the official Crossville weather is measured 2,000 broken, 6,000 overcast, visibility 10 miles—this was at 1700—the man at the radio station, however, said that the clouds have moved in, the visibility is dropping, it's raining pretty hard now at Crossville.

61B—Atlanta Center, 61 Bravo, over.

ATL—November 61 Bravo, stand by one.

61B—Bravo.

TIME: 1122

ATL—November 61 Bravo, go ahead.

61B—Uh, 61 Bravo, over.

ATL—November 61 Bravo, uh, did you get the Crossville weather?

61B—Six-One Bravo, give it to me.

ATL—Roger, November 61 Bravo, Crossville weather measured 2,000 broken, 6,000 overcast, visibility 10; this is the official weather, however, uh, the man at the radio station said the weather to the west is moving in and it's raining pretty hard, visibility is dropping.

61B—Six-One Bravo, if I can get in there, I'd like to get in there.

ATL—Roger, November 61 Bravo is cleared to the, uh, Crossville VOR via direct; descend and maintain 7,000.

61B—Uh, cleared to the Crossville VOR, er, uh, descend and maintain 7,000; leaving 10 for seven now, over.

ATL—November 61 Bravo, roger.

TIME: 1124

ATL—November 61 Bravo, what is your, uh, altitude now?

61B—Six-One Bravo, 8,000.

ATL—Roger.

ATL—November 61 Bravo, Atlanta.

61B—Uh, this is 61 Bravo, I'm, uh, losing power, I've got ice in my carburetors (unintelligible), over.

ATL—November 61 Bravo, roger, understand you are losing power, you've got icing in your carburetors, uh, right now exactly, uh, two miles south of the Crossville Airport. I have no traffic below you, uh, continue descent, and I'll keep you advised.

TIME: 1125

61B—Six-One Bravo, leaving seven now.

ATL—Roger.

UN—Uh, roger, 61 Bravo, you're probably taking ice on the props—why don't you cycle? We're another 560.

UN—Six-One Bravo, uh, this is another aircraft, you probably have ice on your propellers. If you'll cycle the propellers, you should throw it off.

61B—Six-One Bravo, roger, and I have trouble receiving you.

ATL—November 61 Bravo, roger, the aircraft advised you probably have ice on the propellers. You might, uh, try cycling your propellers, and you're cleared for an approach to the Crossville Airport; no aircraft below you.

TIME: 1126

UN—Open your alternate-air doors, will give you the equivalent of carburetor heat, and then also cycle your propellers, run 'em up to full rpm slowly, and then bring 'em back to low rpm, then bring 'em back to cruise, should throw your ice off.

ATL—November 61 Bravo, your radar position now four miles to five miles south heading of about 170 from the Crossville Airport.

ATL—November 2761 Bravo, Atlanta, do you read the Center now?

61B—Six-One Bravo, I read you now, the ice got off my antenna, over.

ATL—November 61 Bravo, understand that you have lost your ice, is that affirmative?
61B—Uh, broke off, now I'm in the rain.

TIME: 1127
ATL—November 61 Bravo, roger, your position now one mile north of the Crossville VOR, and what is your present altitude?
61B—Six-One Bravo, present altitude 6,000 feet, and I would like to be cleared for the approach.
ATL—November 61 Bravo, roger, you are cleared for an approach to the Crossville Airport, contact Crossville Radio 123.6, and I have lost radar contact.

TIME: 1128
61B—123.6, thank you.

Crossville Flight Service Station is not equipped with a recording device, but the statement of the specialist on duty at the time provides an accurate record of the sequence of events that followed. "At 1127," he said, "Atlanta Center advised Knoxville Approach Control and me that N2761B was over Crossville and had carburetor ice and would be landing at the Crossville Airport. The controller advised Knoxville Approach Control and Crossville Radio that N2761B was estimating the Crossville VOR at 1128 and would be at 7,000 feet msl. At 1129, N2761B contacted Crossville Radio and advised that he was at 6,000 feet msl and was over the Crossville VOR. This information was forwarded to Knoxville Approach Control, and a clearance was issued for an approach to the Crossville Airport. N2761B was given the following clearance: 'ATC clears N2761B for an approach to the Crossville Airport.' N2761B acknowledged that he was cleared for an approach to the Crossville Airport. Then I gave N2761B the Crossville weather and airport advisory. N2761B acknowledged, and that was my last contact with the aircraft. Further efforts to contact the aircraft were fruitless. At 1142, I advised Knoxville Approach Control that I lost contact with the aircraft. At approximately 1145, Atlanta Center advised that the aircraft was considered overdue."

The wreckage of the Aero Commander was located the following morning a quarter mile southeast of the Crossville VOR. The VOR itself is at the very peak of Hinch Mountain, 3,048 feet above sea level. The airplane hit the south slope of the mountain at approximately 2,950 feet msl. The form of the leading edge of each wing and the position of the engines in the ground indicated an impact angle of almost 90 degrees, which, due to the steep slope of the mountain, meant that the Commander was about 45 degrees nose-down from level flight when it hit. This was corroborated by the fact that the entire wreckage was contained at the point of impact: damage was confined to extreme fore-and-aft compression, with no wreckage scatter path. The trees and heavy brush surrounding the scene were undamaged, and only the branches almost directly above the aircraft were broken. There was no fire; all occupants died at impact.

A detailed examination of the cockpit and instrumentation showed the aircraft to be in landing configuration with gear extended and flaps one-quarter down. The rate-of-climb indicator read 1,800 fpm down, and the altimeter was impact-jammed at 5,070 feet. Because of the wide difference between altimeter reading and the elevation of the crash site the pitot-static system was given special attention. Although the tubing involved in this system suffered considerable damage at several points, no moisture or indication of preimpact damage was found. The altimeter had been torn from its mountings by the force of the crash and displayed considerable external damage. Normally postcrash instrument readings are given prime consideration: in this case, however, the hands of the instrument were frozen in position by internal damage and not impinged to the face of the instrument, which made the accuracy of the reading highly questionable. No indication of an inflight structural or control failure could be found. The powerplants were disassembled and examined, but no evidence of failure or malfunction of either engine was apparent. The propeller blades displayed distortion and bending opposite to normal rotation, indicating that at least some power was developing when the airplane hit the ground.

The wreckage examination included a search for charts and approach plates. A book of approach plates was found, and it contained the plate for a VOR 1 instrument procedure to the Crossville Airport. The elevation of the Crossville VOR is clearly depicted on this plate. Also located within the cabin were enroute low-altitude charts L-13, L-14, L-21, and L-22. Chart L-21 depicts Victor 51 from Livingston to Crossville but does not include the Crossville VOR. Chart L-14 shows the Crossville VOR but does not indicate its elevation. Both charts, however, show the Crossville Airport and its elevation of 2,023 feet.

While the pilot of the Aero Commander held a commercial certificate with single-engine, multiengine, and instrument ratings, and had a total flight time of 889 hours, he had logged only 37 hours in actual IFR conditions. In light of the fact that his weather briefing indicated penetration of an extremely active and fast-moving cold front associated with forecasts of icing and turbulence, it might have been wiser for the pilot to have delayed or changed the course of his flight.

Less than one hour after takeoff, he first mentioned the icing that was to trouble him throughout the flight. After climbing from 9,000 to 11,000 feet, the pilot reported that he was between layers, the icing problem temporarily relieved; shortly thereafter, though, the pilot struggled for 10 minutes with a relatively simple clearance, again reported ice-related troubles, and abruptly requested a landing at Crossville. From this moment on, the pilot appeared to have been obsessed with a single purpose: get the airplane on the ground at the nearest possible field. No alternative was considered, in spite of the possibility of heavier icing at a lower altitude and of the report of deteriorating weather at Crossville and despite the fact that the terrain beyond Crossville drops rapidly and that points along his route were reported improving weather, with his destination, Atlanta, reporting VFR conditions.

As the Commander descended through 8,000 feet, the harassed pilot said that he was losing power due to carburetor ice. This power loss was more likely due to air intake icing, and there also may have been rime ice accumulating on the propeller blades. At 1125, another Aero Commander suggested a means of clearing the prop of ice and advised use of the alternate-air doors to cure the intake-icing problem. At 1126, the pilot reported that he was flying in rain, with the ice clearing from his aircraft. At this point, his radar position was one mile north of the VOR, altitude 6,000 feet. The Aero Commander was now in an excellent position to cross the VOR and make a procedure turn at the recommended altitude of 5,000 feet. From there, the pilot could cross the VOR inbound at 4,500 feet to complete the normal approach procedure to the MDA of 2,340 feet, or 459 feet agl. The postcrash examination of the frequency settings on the various communications and navigation radios showed the proper selection of frequencies and bearings. The location of the wreckage and its heading at impact indicate the proper inbound course to the VOR after procedure turn. Why hadn't the approach terminated in a safe landing at Crossville?

Although relieved of the ice that had plagued him, the pilot still carried the memory of surging engines and vibrating, ice-laden props into his final approach. All he wanted to do was to reach the safety of the runway as rapidly as possible—a runway that remained unseen only a few miles ahead, hidden by a layer of clouds 2,000 feet above it. A quick check of the charts or approach plate told him that the airport elevation was 2,023 feet. At an altitude of about 4,000 feet, he should drop out of the blinding clouds. After turning toward the field and nearing the VOR caution was perhaps replaced by a desperate urge to get ground contact, to assure landing on the first attempt.

In the final moments of the steep descent the distracted pilot probably did not monitor his rapidly diminishing altitude, anticipating only the moment when he would clear the clouds. From the time that the pilot of 61B made the decision to land at Crossville to the moment of impact, little more than 10 minutes passed. Under the pressure of ice, partial power loss, and an IFR approach to an unfamiliar field, it is doubtful that he had time to study the approach procedure closely. In all probability he never realized that the VOR was more than 1,000 feet above the runway, blocking the path of an approach that he would never complete.—W. L. T.

The pilot was obsessed with the idea of landing, so, although he apparently could have continued to Atlanta, only 45 minutes away, once he got into rain and below the freezing level, he discarded that possibility. At the last moment he must have lost control of the airplane: the 45-degree nose-down attitude suggests that he may have caught sight of the ridge just before impact, pulled up, and spun into the ground.

He was undone by fear. He was an inexperienced instrument pilot and was uneasy about the coming trip, as his frequent visits to the weather desk days in advance suggest. Perhaps he had never flown in icing conditions. The decision to take off, made under the pressure of established plans and a commitment to his companions, may have been unwise. But there was nothing in the circumstances of the flight that warranted its end except the pilot's own state of mind. Perhaps if a controller had mentioned to the pilot that, if he was out of icing conditions at 6,000 feet, he might want to continue to Atlanta, where the weather was fine, then the poor man might have realized his alternative. Most likely he would not: in the terrifying isolation of the cockpit, he must have felt that the only safety and solace lay in escape from the gray emptiness that oppressed him. Keeping fear at bay in the alien and unreal surroundings of IFR flight is a trick that pilots must learn; this pilot did not learn it in time.—P. G.

6.

PANIC

On Saturday, May 4, 1969 a 42-year-old executive left Teterboro, New Jersey on a long anticipated cross-country flying vacation to Los Angeles. He was accompanied by his wife, their four children (all under five), and his father-in-law. The pilot had been planning the trip in great detail for some time, going so far as to meet with his flight instructor for advice and assistance. The aircraft—a Beech C33 Debonair N2737T—had been purchased new by him and had dual VHF transceivers, dual VOR/ILS with glideslope receiver, ADF, DME, and transponder. The aircraft was also fitted with long-range fuel tanks, autopilot, and oxygen system; the only anti-ice equipment, however, was a heated pitot tube. The pilot had 300 total hours on a commercial license with single-engine land and instrument ratings, and 300 total hours, of which 50 had been logged during the preceding 90 days.

Because the flight was to be conducted during the daylight hours only, several overnight stops were planned. The first was at Lexington, Kentucky. The following day, the aircraft arrived at Wichita, Kansas, where the group stayed overnight before attempting the next leg, which would take them over mountainous terrain. At 0858 on the morning of May 6, N2737T left Wichita for Farmington, New Mexico. Prior to departure, the pilot received a full weather briefing by telephone from the Wichita FSS. The weather forecast indicated that he could expect to encounter IFR conditions with possible icing in the clouds over the Rocky Mountains. The destination airport was expected to be VFR by the estimated time of arrival. The pilot then filed an IFR flight plan via V73 to Hutchinson, Kansas; V10 to Lamar, Colorado; and V210 to Farmington. The initial cruising altitude was 8,000 feet, with a request for 15,000 passing Lamar. After takeoff from Wichita N2737T maintained normal radio contacts with Kansas City Center. The flight proceeded without reported incident and at 1040 was approaching the Lamar VOR at 8,000 feet, precisely on schedule. At this point, air traffic control was passed to Denver Center. The following is a transcript of the relevant radio contacts with the pilot (37T = Beechcraft N2737T; DEN = Denver Center Radar; FMN = Farmington FSS and Tower; * denotes contacts between ground stations on direct land line).

TIME: 1040
DEN—Beechcraft 2732 [sic] Tango, Denver Center, 8,000, report Lamar.
37T—Beech 2737 Tango, I will.

DEN—Beech 37 Tango, squawk 1100, ident.

37T—Three-Seven Tango, ident.

DEN—Three-Seven Tango in radar contact now nine miles east of Lamar.

37T—Three-Seven Tango, roger, thank you.

TIME: 1051

37T—Denver Center, Beechcraft 37 Tango.

DEN—Three-Seven Tango, go ahead.

37T—Yeah, keep in mind I'm on my way to a higher altitude, sir.

DEN—Yes, sir, I haven't forgotten you. You can expect higher in, ah, 15 miles.

37T—Three-Seven Tango, roger, thank you.

TIME: 1055

37T—Denver Center, 37 Tango. When you get an opportunity, check Farmington weather, please.

DEN—The 1600 Farmington weather is, ah, estimated ceiling 2,000 overcast, visibility 10, and light rain. Temperature 40, wind 210 degrees at seven knots.

37T—Three-Seven Tango. That was, ah, be, ah, kind enough to check out a good alternate for me, because I was given way above VFR, ah, upon my, ah, estimated time of arrival at Farmington, things seem to have changed.

DEN—Ah, 37 Tango, you'll have to go to La Junta Radio for, ah, any weather or alternates. We don't have that information.

TIME: 1057

37T—Three-Seven Tango. I'll check with you again upon arriving at Farmington.

DEN—Three-Seven Tango, ah, roger, you can contact Farmington Radio when you're a little ways out, ah, to get the same weather, more current weather.

37T—Three-Seven Tango. Thank you.

DEN—Beechcraft 237 Tango, climb and maintain, ah, 15,000, over.

37T—Three-Seven Tango. We'll, ah, climb to and maintain 15,000.

TIME: 1120

DEN—Three-Seven Tango, roger, report reaching 15,000.

37T—. . . 'll report reaching 15.

37T—Denver Center, 37 Tango. Reporting level at 15,000.

DEN—Three-Seven Tango, roger.

DEN—Beech 2737 Tango, contact Denver Center on 126.1.

TIME: 1140

37T—That's 126.1, thank you, good day. Three-Seven Tango.

DEN—Good day, sir.

37T—Denver Center, 37 Tango, maintaining 15,000.

DEN—Ah, Beechcraft 27 [sic] Tango, Denver Center, ident.

DEN—Beechcraft 37 Tango, Alamosa altimeter 30.00.

37T—Three zero zero zero, 37 Tango, thank you.

37T—Denver Center, 37 Tango, could you get a good weather check on, ah, Farmington, New Mexico, weather, please?

DEN—Beechcraft 37 Tango, ah, Farmington one-hour-old weather, measured ceiling 1,100 overcast, visibility six miles, light rain and fog.

37T—Is there a nearby airport that, ah, would have any kind of VFR at this time?

DEN—Ah, Beechcraft 37 Tango, say again?

37T—I say, is there a nearby airport that would be, ah, VFR at this time, because it's quite apparent I'm not going to be able to make it at, ah, Farmington. By the time I get there, it'll probably be way down.

DEN—Beechcraft 37 Tango, stand by; I'll check on it.

TIME: 1153

*FMN—Farmington Tower.

*DEN—Say, is, ah, your ceiling and visibility still about, ah, 1,100 overcast, six miles?

*FMN—Just a second, let me check it here.

37T—Denver Center, 37 Tango. How about Alamosa Municipal?

DEN—Ah, Beechcraft 37 Tango, I'm checking with Farmington Tower now, they advise it is breaking. They're taking another reading on the, ah, ceiling and visibility.

*FMN—Ah, Farmington Radio.

*DEN—Ah, Radio?

*FMN—Yeah.

*DEN—Ah, is the ceiling and visibility still, ah, 1,100 and six?

*FMN—No. We got, ah, 400 scattered now, measured 1,400 overcast, 15 miles visibility, very light rain.

*DEN—Very light rain, ah, it's breaking up?

*FMN—Yes, it is, it's breaking up, ah, fairly rapidly right now.

*DEN—Ah, ah, would you say, ah, in an hour; what would you guess?

*FMN—Oh, I'd imagine it'll probably, ah, it'll be about 2,000, I imagine.

*DEN—About 2,000 overcast?

*FMN—Ah, yes, uhuh.

*DEN—Okay, thank ya much.

DEN—Bonanza 37 Tango, ah, Farmington advises, ah, that right now they have 400 scattered, measured ceiling 1,400 overcast, visibility 15, and very light rain. He says it's breaking rapidly and expect in one hour to have about 2,000 overcast.

37T—Three-Seven Tango, right, ah, fine. Then I'll push right on to Farmington. Thank you very much, 37 Tango.

DEN—Roger, and, if you can't make it into Farmington, ah, Cortez and Durango are both reporting about the same, ah, 6,000 ceiling with 30 miles visibility.

37T—Three-Seven Tango. Fine, I'll try it that way then, thank you very much.

TIME: 1206

37T—Denver Center, Denver Center, emergency. Denver Center, emergency. Three-Seven Tango, emergency.

DEN—November 37 Tango, this is Denver Center, say again.

37T—Emergency—I've all of a sudden lost my airspeed, I don't know why.

DEN—Ah, November 37 Tango, roger, you still at 15?

37T—Ah, how low can I go?

DEN—Are you at 15,000, sir?

37T—Denver Center, how low can I go at this area?

DEN—Ah, one, ah, four thousand, 14,000.

TIME: 1207

DEN—November 37 Tango, what is your present heading, sir?

37T—Can you vector me around to a lower area?

DEN—November 37 Tango, ah, what is your present heading, sir?

37T—Can you vector me around to a lower area?

DEN—November 37 Tango, ah, what is your present heading, sir?

37T—Can you vector me to a lower area, sir? I'm at 14,300.

DEN—Roger, turn south heading 180, turn south heading 180.

37T—Center, are you reading me, sir? I'm 14,000 going down.

DEN—Ah, roger, are you heading to the south? Are you heading 180 now, sir?

37T—Denver Center, can you help me?

DEN—November 37 Tango . . .

37T—Denver Center, can you help me? Emergency.

DEN—November 37 Tango, affirmative, ah, try to ease off on your, ah, controls. Keep your throttle forward, keep your throttle forward. Maintain as much altitude as possible, please.

37T—I'll try to maintain 14,000 if I can, sir. My airspeed is way down.

DEN—Ah, your airspeed working now?

37T—Yes, sir (unintelligible).

DEN—November 37 Tango, do you have any airspeed at all?

37T—Any airspeed what, sir?

DEN—Ah, is your airspeed indicator operating?

37T—Yes, sir, it is.

DEN—You say that is affirmative, affirmative?

37T— . . . firm, affirm, yes. I'm using my static air source, my second static air source. I'm 13.7, sir.

DEN—Ah, roger 13.7 and you're heading south, is that correct, sir?

37T—I'm heading 240.

DEN—Is that 240?

37T—Ah, 300.

DEN—Two zero zero, is that correct?

37T—(Unintelligible) . . . 'ee zero zero, yes, sir, I'm going to get back on course now, sir.

DEN—Ah, roger, and what is your altitude, sir?

37T—Thirteen five hundred.

DEN—Ah, keep your airspeed up as high as possible, keep your airspeed up as high as possible, please, November 37 Tango. Do you have your throttle advanced, ah, to maintain as much speed as possible?

37T—Yes, sir, I have.

TIME: 1209

DEN—Ah, what is your altitude now, sir?

37T—Altitude 13,200, airspeed 50 miles an hour, going down to zero and back up again.

DEN—You say you're going to try to go back up again?

37T—I'm trying to hold it up.

DEN—Ah, 37 Tango, are you in VFR conditions now?

37T— . . . gative, I'm socked in completely.

*FMN—Farmington Radio.

*DEN—Farmington Radio, we've got an emergency out here, just, ah, east of the field, ah (unintelligible), your DF on? He's transmitting on, ah, 126.1.

*FMN—One twenty-six, ah, one?

*DEN—Yeah, he's losing altitude. He's, ah, he's lost his airspeed, he's going down.

*FMN—Oh, oh, 126.1, huh?

*DEN—Yeah.

*FMN—Don't have that on this, ah . . .

*DEN—You don't?

*FMN—No, ah, how far east is he?

*DEN—Ah, I can't talk any more, there's too much . . .

TIME: 1210

37T— . . . transponder, sir?

DEN—November 37 Tango, say again?

37T—Do you have me on transponder? Do you have me on radar?

DEN—Ah, negative, sir, radar contact lost. What is your altitude now, sir?

37T—Thirteen thousand, sir.

DEN—Thirteen thousand, and do you have your throttle forward, is that correct?

37T— . . . sir, I'm heading 290 at this time.

DEN—Roger, heading 290. What is your altitude now?

37T—Thirteen thousand, sir.

DEN—Roger, 13,000. Are you in the clear at this time?

37T— . . . gative, I'm encountering icing on my windshield.

TIME: 1211

*DEN—Ah, Farmington?

*FMN—Yeah.

*DEN—Do you have any reports of, ah, weather out east of the field?

*FMN—Ah, no, it looks pretty good east of the field out there. I can see, oh, 20, 25 miles, ah, northeast. There's some fog but, ah, but looks in general, looks real good.

*DEN—What's the bases?

*FMN—Oh, the bases about 400 scattered and, ah, then we got some higher cirrus above it.

*DEN—Uhuh, about where does it break off? Ah, do you see any type weather to the east of that?

*FMN—Ah, no, it looks, ah, real good, ah, all the way east and to the southeast of us here. A little weather to the northeast, ah, quite a ways up there, but as a rule it looks real good to the east.

*DEN—Okay.

*FMN—Ah, Center?

*DEN—Yes?

*FMN—How far east is he?

*DEN—I don't know, ah, he's pretty panicky right now, and I can't get to the ra . . .

*FMN—Ah, okay, real good.

DEN—Ah, November 37 Tango, do you have any visibility out of your left side or your right side?

37T—No visibility on any place.

DEN—November 37 Tango, can you open the window a little bit? Could you open the window a little bit?

37T—I can, sir, why? No, I can see out the window, but I'm socked in, oh, I can see visibility now. I can see (unintelligible—15 miles in here or 15 mountains here), sir. I had a little visibility; I can see mountains.

DEN—November 37 Tango, what is your airspeed?

37T—Airspeed keeps going up and down, 40 and down, I'm going down, sir.

TIME: 1212

DEN—Ah, what is your heading now?

37T—Heading is 280.

DEN—Roger, holding heading 280. What is your altitude now, sir?

37T—Twelve thousand five hundred, going up.

DEN—Roger, 12,500, and what does your airspeed indicator read?

37T—It indicates about 50, and I'm going up again, ah, starting to go up again, sir.

DEN—Roger, you're starting to go up again, and, ah, can, do you have any visibility at all?

37T—No visibility now, I lost it again . . .

DEN—No visibility, roger.

37T—Can I get out of this below, sir?

DEN—Ah, 37 Tango, are you in the clear, are you in the clear?

37T—Negative, sir, I see a little faint, but not much, very hazy.

37T—Is there any clearing around here at all, sir, that I can head for?

DEN—Are you still heading, ah, to the west? Are you still heading to the west at this time?

37T—Two six zero now, I do see some mountains, sir, but very hazy.

DEN—Try to stay in the clear, if possible; ah, do you have any visibility at all?

37T—Keep going in and out of my visibility, sir.

DEN—Roger, remain VFR conditions, if, ah, possible . . . main VFR if possible, over.

37T—I'll try, sir.

TIME: 1213

DEN—Roger, continue to the west if you can and remain VFR, remain VFR. Head to the west as long as you can.

*FMN—Farmington Radio.

*DEN—Yeah, what's your visibility, ah, at the field?

*FMN—At the field, ah, we've got better than 15 miles.

*DEN—Is it that way about all quadrants?

*FMN—Yes, right.

*DEN—Okay, thank you.

*FMN—Uhuh.

DEN—November 37 Tango, what is your present heading, sir? November 37 Tango, what is your present heading? November 37 Tango, what is your present altitude, over?

TIME: 1215

37T—I can't keep it too long (unintelligible).

DEN—Ah, say again altitude, say your altitude, November 37 Tango. Maintain as much altitude as possible. If you are VFR, maintain VFR, November 37 Tango, Denver Center, over.

No further contact could be made with N2737T. An extensive search was conducted over the route of flight for several weeks, but no trace of the aircraft or its occupants was found. The wreckage was finally located on August 22, 1969—more than three months later—in a dense stand of tall trees at an elevation of 11,000 feet, just slightly north of the flight-plan course. Prior to impact, the aircraft had sheared off 10 trees in a relatively shallow angle of descent. Sections of both wings had separated on contact with the trees and were found between the first broken tree and the main body of wreckage. Postimpact fire destroyed the fuselage from the firewall aft to the baggage compartment and all instrument readings and evidence of control settings. No signs of power, structural, or control malfunction prior to impact could be found, however. All occupants were killed in the crash.

The final flight of N2737T presents a tragic example of the often fragile restraints that separate a usually cautious and competent airman from the mindless reflexes of sudden panic. It also offers a vivid picture of how otherwise minor circumstances can combine to bring disaster. From the information available it is obvious that the pilot took unusual care in planning each segment of his intended trip, even to the point of seeking advice from his instructor. Perhaps, due to the extent to which the course had been preplanned, he was reluctant to alter his plans by considering another route when the weather and terrain dictated an alternate course to be more prudent.

As he prepared to depart Wichita on what would so tragically become the last leg of the flight, the pilot received a weather briefing that indicated IFR conditions with possible icing over the higher mountains along his route. Since the destination airport at Farmington is not equipped with an instrument approach, the pilot was particularly careful to see that the forecast was for weather well above VFR at the estimated time of arrival. He then accepted this forecast without planning for an alternate airport should the weather fail to clear. There is no evidence that he checked the weather for—or even considered—a more frequently used route from Wichita to Los Angeles south of the higher mountains and poor weather, via airways to Amarillo, Texas, and Albuquerque, New Mexico.

The recorded radio contacts with the flight after takeoff from Wichita showed the competence of the pilot as he responded to the routine demands of IFR flight. His alertness was evident when at 1051 he reminded Center of the higher altitude required on the airway passing the Lamar VOR. At 1055, the pilot first learned that the Farmington weather was considerably less than forecast. At this time, the voice recordings show that a definite change in the pitch and composure of the pilot's voice occurred. His request for information on a possible alternate airport was made in a nearly pleading tone and gave evidence of his deep concern. In the following exchanges, the pilot had some difficulty in expressing himself. However, after being assured by the controller that the Farmington weather was rapidly improving, the pilot was obviously relieved, and his voice again returned to normal.

At 1206, the first indication of serious trouble came as the pilot declared an emergency. At this point his voice was shrill and barely readable. He reported failure of the airspeed indicator and then repeatedly pleaded to be vectored toward lower terrain; although the controller immediately issued a new heading, the pilot repeated the request several times as he became increasingly excited and nearly incoherent. The heading that he so desperately demanded was never acknowledged as the aircraft continued westbound.

Less than three minutes later, the pilot reported that he had switched to the alternate air source and indicated that the airspeed was again functional. However, a few minutes after this he gave his airspeed as 50 mph during a climb. There is little doubt that the airspeed indicator was still inoperative.

From the first report of the emergency until the final contact 12 minutes later the pilot was totally obsessed with the erratic airspeed readings and the relentless loss of altitude. A basic maneuver such as the lifesaving 180-degree turn never seemed to be a consideration.

From all the evidence available the only actual problem facing the pilot of N2737T was the failure of the airspeed indicator, most likely due to pitot-tube icing. This is further confirmed by the continued false readings after selection of the alternate static source. The possibility of static-source icing is remote.

Seeing the unbelievably low readings on the fluctuating airspeed needle, the pilot was faced with a situation that he had never experienced during instrument training. He reacted to the airspeed in the only way he knew—by lowering the nose. When the expected correction did not occur, he first experienced fear, which quickly turned to total panic. The continued urging of the controller to keep the throttle forward and maintain airspeed only gave validity to his instincts, and he continued to descend in an attempt to regain the faltering airspeed. Descending through 13,000 feet, more pressure was placed on the nearly terrified pilot, with the added problem of windshield ice. Occasional fleeting glimpses of the ground as the aircraft flashed in and out of the ragged ceiling encouraged continuing descent in a fruitless search for clear skies.

Loss of the airspeed indicator, even in solid IFR conditions, presents only a minor inconvenience for the experienced pilot. Normal procedure in this event is to disregard the instrument completely and to rely solely on power setting and the altitude instruments to maintain level flight. Sadly, however, because of limited time and the lack of a requirement to do so, most pilots are never exposed to the failure of the airspeed indicator during training for any level of pilot certificate, although reliance on this basic instrument is stressed for all pilot ratings.

The degree to which this minor failure affected the pilot of N2737T, who had a reputation as an extremely alert and thorough airman, is evidenced by his failure to consider the most obvious solution to his dilemma: at the precise moment that he reported his emergency the radar position of N2737T was only 10 miles west of Alamosa, Colorado, located in a broad valley at an elevation of 7,535 feet. Alamosa at that time was reporting an estimated 4,000 feet broken, 10,000 feet overcast, visibility 40 miles. This meant that the ceiling was at a flight altitude over 12,500 feet, and clear skies were 2,500 feet directly below the stricken Debonair. It is ironic that this crash may have been the result of the failure of the only anti-icing device installed on the aircraft—the heated pitot tube. Or perhaps the pilot simply neglected to turn it on during that fateful leg of his final flight.—W. L. T.

There is no requirement that a controller be a pilot, though many are, and many pilots in fact resent attempts by controllers to tell them how to fly their airplanes. In this harrowing instance, however, the pilot himself had forgotten how to fly, and the controller had to take over for him. Unfortunately, this controller did not come up with the right answers. It's easy to second-guess in retrospect but difficult to know what to do for a panicky pilot with whom one's only contact is by radio. The pilot should have observed the sudden low reading of his airspeed indicator with suspicion; verified his attitude and altitude by his other instruments; checked that the pitot heat was on and that the pitot-heat circuit breaker had not popped; gone to his alternate static source; and then, if the airspeed indicator refused to cooperate, ignored it and continued flying toward Farmington or checked his charts for an alternate place to land, such as Alamosa. So far as we know, he took only one of these steps, that of selecting the alternate static source. Most likely the problem was caused by ice in the pitot head; at any rate, it was trivial, except that for this novice instrument pilot, far from home in a crowded airplane in uncertain weather at an altitude probably higher than he had ever been before, it broke his composure. The controller's ill-advised suggestion to maintain airspeed by pushing the nose down merely supported the pilot's irrational idea that the airplane rather than the airspeed measuring system was at fault. The controller's suggestion that the pilot try to remain VFR was also unwise: the best, the only thing for him to do under the circumstances would have been to continue along the airway, maintain altitude, and try to pull himself together. *The pilot's panic persuaded the controller that there was a material emergency when there was in fact none.*

This account is particularly painful to read; it is so frustrating, so unnecessary, and yet one cannot help sharing the pilot's helpless terror. The story underscores the absolute necessity of controlling fear; when fear enters, it is like a solvent that weakens reason and fragments perceptions; panic is like a flood that carries everything before it.—P. G.

7.

"TRYING TO GET BACK TO VFR . . ."

At approximately 1047 on the morning of August 30, 1969, the pilot of Piper Twin Comanche N8391Y departed Richmond, Virginia on a VFR flight plan to Buffalo, New York. Aboard were two passengers, one of whom was in a full body cast and was being transported to a Buffalo hospital for treatment. The pilot, age 46, was president of his own manufacturing company and had donated his time and use of the aircraft for the mercy flight. No radio contacts were reported until 1222, when the flight contacted Cleveland Center. The following is a transcript of those recorded transmissions (91Y = Piper Twin Comanche N8391Y; ZOB = Cleveland Center, Erie Sector Radar; ERI = Erie Tower/ Approach Control; BFD = Bradford Flight Service Station; 550 = Allegheny Airlines Flight 550; UNK = unidentified transmissions; * denotes direct-line communication between ground stations).

TIME: 1222
91Y—Cleveland Center, 8391 Yankee.
ZOB—Eight-Three-Niner-One Yankee, Cleveland, stand by.

TIME: 1225
91Y—Cleveland Center, 8391 Yankee.
ZOB—Eight-Three-Niner-One Yankee, Cleveland, go ahead.
91Y—Approximately 30 miles southeast of Bradford VOR, descending, leaving 7,000, transponder-equipped, landing Buffalo.
ZOB—Eight-Three-Niner-One Yankee, ah, Cleveland, say again, please.
91Y—Eight-Three-Niner-One Yankee. Thirty miles southeast of Bradford VOR, landing Buffalo, leaving level 7,000 now.
ZOB—Niner-One Yankee, are you VFR requesting advisories?
91Y—Affirm.
ZOB—Eight-Three-Niner-One Yankee, ident code 0600.

TIME: 1226
91Y—Niner-One Yankee. Squawk ident on 0600.
ZOB—Eight-Three-Niner-One Yankee, ah, we have you radar contact, sir, and for advisories contact Cleveland Center 133.95.
91Y—Three three five?
ZOB—One three three niner five.
91Y—Don't go that high. I go 135.9 [sic].

ZOB—Roger, remain this frequency, sir.
ZOB—And 91 Yankee, say your type of aircraft, sir.
91Y—A Twin Comanche.

TIME: 1227
ZOB—And 91 Yankee, what will be your cruising altitude, sir?
91Y—I'm trying to get down where I can get back to VFR conditions.
91Y—Center, I have a VOR of 116.4 for Buffalo. Is that correct? I'm not picking it up.
ZOB—Niner-One Yankee, if you are calling, I was off the frequency. Say again, please.

TIME: 1228
91Y—I have a VOR setting of 16.4 for Buffalo. Is that correct?
ZOB—Roger, Buffalo frequency is 116.4.
91Y—I'm not picking it up, that's why I wondered.

TIME: 1229
91Y—Leaving level 6,000.
ZOB—Niner-One Yankee, roger.

TIME: 1230
91Y—Niner-One Yankee, and I'm no longer VFR.
ZOB—And 91 Yankee, you're garbled; ah, I got your report out of 6,000 descending VFR, is that affirmative?
91Y—Affirm, trying to descend VFR.
ZOB—Roger.
91Y—However, I'm in the clouds.
*ZOB—Erie Approach, Cleveland.

TIME: 1231
*ERI—Erie.
*ZOB—Okay, we have an aircraft descending in the clouds [in background 91 Yankee reports level at 5,500 feet in the clouds], he's at, ah, 5,500 in the clouds, he's, ah, he's supposed to be VFR. He's presently 10 miles southeast of Bradford—is your inbound flight out of six or five?
*ERI—He's out of six, but he's in the same vicinity going right down to four.
*ZOB—Yeah, he's south . . .
ZOB—Niner-One Yankee, this is Cleveland Center, stand by, please.
91Y—Niner-One Yankee, standing by.

TIME: 1232
ZOB—Niner-One Yankee, are you capable of and qualified for IFR flight, sir?
91Y—(unintelligible) . . . what's the final . . . (unintelligible) . . . we're dropping 500 feet a minute.
ZOB—Niner-One Yankee, be advised an aircraft on approach to Bradford, just out of 5,000, ah, northeastbound, last reported eight miles southwest of the

Bradford VOR, ah, you're presently about seven miles southeast of the Bradford VOR.

91Y—Niner-One Yankee. I'm climbing out of this to, ah . . . I'm not VFR at 5,500.

TIME: 1235

91Y—Niner-One Yankee. I'm level at 5,500—what should I do?

ZOB—Niner-One Yankee, are you in VFR conditions, sir?

91Y—I can't see the ground; I'm above the clouds, though, I'm running into some higher clouds now.

ZOB—Niner-one Yankee, are you qualified and capable for IFR flight?

91Y—No, sir, I'm not checked out IFR.

ZOB—Roger, can you maintain VFR above the clouds and establish ground contact southeast of Bradford?

91Y—I can just stay above the clouds. There's no problem there. I want to get down through and see what I'm doing.

TIME: 1236

91Y—I'm past the Bradford VOR right now.

ZOB—Roger, I have you in radar contact, sir. You are about five miles south of the Bradford VOR. The Buffalo weather is sky partially obscured, estimated 8,000 broken, visibility two, haze and smoke.

*ZOB—Erie Approach, Cleveland.

*ERI—Erie.

*ZOB—Okay, ah, the aircraft called about now, he's at 5,500 on top, but he can't see the ground, so he's going to run into some higher clouds. I don't know what he's going to do.

TIME: 1237

*ERI—Okay, we don't have anything right now at Bradford.

*ZOB—Yeah, well . . .

*ERI—Which way is he going? Is he . . .

*ZOB—Northbound, he's presently five miles north of Bradford VOR.

*ERI—Five miles north of Bradford, eh?

*ZOB—Yeah.

*ERI—Okay, good luck.

*ZOB—Thanks, I need it.

*ERI—All right, you got Bradford's weather, don't you?

*ZOB—Yeah, it's sky partially obscured, visibility three.

ZOB—Niner-One Yankee, say again, please.

91Y—Did you say the final . . . the visi . . . the altimeter at Buffalo was 2,000 feet?

ZOB—Ah, no, the cloud layer at Buffalo is estimated 8,000 broken, sir.

TIME: 1238

550—Center from Allegheny, if that light aircraft is landing at Buffalo, you might advise him that the visibility is, ah, variable and very restricted at times.

ZOB—Allegheny 550, roger, thank you. And 91 Yankee, an Allegheny just came out of Buffalo, advises that the visibility is very restricted at times. We're carrying Rochester as clear, visibility two and one-half with haze, and apparently quite hazy up in that area, sir.

TIME: 1239
91Y—I'm experiencing motor trouble, I gotta let down.
ZOB—Niner-One Yankee, roger, your radar position is, ah, 10 miles north of the Bradford VOR, ah, if you want, you can turn around and proceed to Bradford. I'll advise Erie Approach Control.
*ZOB—Erie, Bradford controller here. Erie, ah, this 91 Yankee north of Bradford, he's turning around, coming back to Bradford. He's in trouble, he's got engine trouble.
*ERI—He's got engine trouble? Okay, you going to turn him over to us or Bradford Radio?
*ZOB—I can turn him over to you—he's at 5,500 feet.
*ERI—Okay, now I'll see. Put him on Bradford Radio—they might be able to help him out that way.

TIME: 1240
*ZOB—Okay, Bradford Radio, what's the frequency?
*ERI—One two three point six.
ZOB—Okay, ah, 8391 Yankee, contact Bradford Radio now, 123.6; they have DF, and they'll be able to give you assistance, sir. I suggest a southerly heading to Bradford.
91Y—Say that again . . . the radio.
91Y—This is 91 Yankee, give me those headings again. I'm in trouble.
ZOB—Niner-One Yankee, understand, ah, I would suggest a southerly heading to Bradford; contact Bradford Radio on 123.6. They have DF available and will be able to assist you, sir.

TIME: 1242
*ZOB—Erie Approach, Cleveland.
*ERI—Go ahead.
*ZOB—Yeah, has that guy called Bradford? It's, ah, 8391 Yankee.

TIME: 1243
*ERI—Yes—hold on a second—Bradford Radio, Erie, if you're not busy.
*ZOB—And, ah, I gave him the frequency a couple of times.
*ERI—Yeah, hold on a second, maybe he's working and can't answer us right now. Bradford Radio, Erie Approach.

TIME: 1250
UNK—North of Bradford, he had eng . . .
UNK—You have a problem?
UNK—Yeah, that . . .
UNK—What about him?

*BFD—Well, we lost him at 1247; we had a fix on him 10 miles north of Bradford, and we had him inbound on a heading. We lost contact with him.

*ZOB—All right, just wait a minute here.

*BFD—Okay, we'll ah, we're checking . . . he's gonna pass just about over the Harri-Emery Airport, which is just north of Bradford about 10 miles . . . we're calling the airport now to see if maybe he dropped down in there.

*ZOB—Oh, we'll get the watch supervisor on here, and what is your current weather?

*BFD—We're clear, ah, sky partially obscured with three miles in haze, about two- to three-tenths obscured.

No further radio contact was made with Twin Comanche N8391Y, but the statement of a ground witness who observed the flight gives a description of the pilot's actions moments before impact: "At approximately 1230, I was in the side yard of my residence when my attention was attracted to an aircraft overhead with an unusual-sounding engine. I observed the twin-engine aircraft at about 300 feet altitude on a southwesterly heading, losing altitude. I observed the right propeller in a feathered position and slowly windmilling. I first thought this aircraft was making a normal 'down the valley' approach to the Harri-Emery Airport, which his position and altitude indicated, and just after passing over my home the landing gear came down. Shortly after the gear was lowered, I noted power application to the left engine. I then observed the aircraft in climbing attitude and I remarked to my father, who was with me, that the plane appeared to be high enough to clear the top of South Hill. I was observing the aircraft through field glasses after he applied power and last observed him over the edge of the ridge at South Hill. It appeared to me that his speed during and after application would not exceed that of a J-3, which are seen in this area daily."

The wreckage of N8391Y was located three miles south of Harri-Emery Airport at Bradford, Pennsylvania at the 1,800 foot level of a 2,100 foot ridge. The slope of the ridge is approximately 45 degrees. All components of the aircraft were found in close proximity to the wreckage. The first point of impact was with two trees about 10 feet above ground level. The aircraft then struck the ground nose-first about 15 feet and came to rest in an inverted position. There was no fire. All occupants were killed at impact.

A detailed investigation revealed that the left engine propeller was in low pitch (high rpm); the right engine propeller blades were in the feathered position. All fuel tanks ruptured on impact, with only residual fuel remaining. The fuel-tank selectors indicated that the left engine was on the left main tank, while the right engine was on the right auxiliary fuel tank. The left propeller blades indicated that little, if any, power was being developed at impact. There was no fuel found in either engine or in their fuel-supply lines. Both engines were removed and sent to the Avco Lycoming factory for inspection and testing. Both powerplants ran normally and developed rated power after replacement of the impact-damaged accessories. No preimpact mechanical failure or malfunction was noted on either engine. The autopsy report noted nothing to indicate physical incapacitation of the pilot prior to the crash.

The flight of N8391Y had evidently proceeded normally until it approached the Bradford area. At this time, the non-instrument-rated pilot found himself on top of an overcast and established radio contact with Cleveland Center Radar. During the initial contacts, the pilot advised Cleveland Center that he was descending through 7,000 feet. At this point, the pilot indicated that he was VFR, and radar contact was established. Only minutes later, at 1227, the pilot said, "I'm trying to get down where I can get back to VFR conditions." Three minutes after this, the pilot reported directly to center that he was no longer VFR. At 1230 both pilot and controller were still unsure about whether the flight was or was not VFR. The pilot said, ". . . trying to descend VFR." Moments later, he added: "However, I'm in the clouds," which arouses doubts about whether the pilot himself understood the definition of VFR. The transcript also makes it obvious that the controller was busy; he may well have missed the harried pilot's ambiguous remark.

At 1236, the pilot was advised that his destination, Buffalo, was below VFR, but no attempt was made to advise him that Bradford was VFR with very light obscuration and three miles' visibility. Finally, at 1239, the troubled flight was turned around for the approach to Bradford. Now the added problem of a failed engine further pressed the pilot. At this time, Cleveland Center and Erie Approach Control seemed to give up completely on any thought of giving continued assistance to the pilot, and he was unceremoniously passed to Bradford Radio for a DF steer to the airport. This action was taken without prior arrangement with Bradford Radio and in spite of the fact that the flight was in radar contact by these two stations, eliminating the need for a DF procedure. Indeed, the already seriously distracted pilot was left to his own devices for some time so that his dilemma could be explained yet one more time to a new and uninformed ground controller.

It is worth noting that the pilot might well have commanded more attention from the controllers had he been less ambiguous about whether he considered himself to be in a genuine emergency. He neither declared the emergency fully and early nor did he go so far as to declare a "Pan" alert. Had the pilot acknowledged that his situation was deteriorating rapidly and that he wanted all the help he could get, he might well have minimized the uncertainty and indecision that prevailed among the controllers, who were unsure as to how to help him.

The pilot never obtained a complete and current weather briefing prior to departure, nor did he check current weather for his route and destination with ground stations during the flight. Finally, he did not request from the radar controller vectors to an area of VFR weather to avoid descending through clouds. With less than two hours' flight time on full tanks, the aircraft's range allowed several alternative airports, including returning to the point of departure. Once in the clouds and in obvious need of sound advice and direction, the pilot should have concentrated on one single goal: to seek VFR conditions as soon as possible.

At the time that the engine failure occurred, the pilot was so intent upon maintaining control of his aircraft during his all-consuming search for ground contact that he instinctively performed the feathering procedure, evidently without attempting to determine the cause of the malfunction. In all probability simply moving the fuel selector from the auxiliary tank to the main would have restored power to the balky powerplant and eased the pilot's mind. During the final moments of the flight, the aircraft was on final approach to Bradford Airport at low altitude and minimum speed, continuing a descent in restricted visibility. With the field in sight, the pilot dropped his landing gear, which only increased drag on the aircraft and required the addition of power. At the last moment, the pilot caught sight of the steep ridge as the aircraft sank lower on his uncertain approach to the strange field. Even with full power the pilot could not overcome the established descent.

One haunting observation in the report remains to cloud any pat conclusion as to the true cause: the absence of fuel in both engines and fuel lines. Had the distressed pilot committed the final error and allowed his one remaining engine to run out of fuel? Or had the residual fuel simply drained from these parts through the ruptures in the lines and tanks after the wreckage came to rest inverted? The latter seems more plausible, but there is no way to know the true answer. Like this pilot, all too many airmen delay uttering the word "Mayday," which is the one way to remove all doubt about the magnitude of a predicament. Help that is available for the mere voicing of those two syllables is allowed to go unused. It is not clear whether this is because of fear of reprimand or because of some inborn resistance to facing up to a bad situation. If a pilot must make the choice, it is always better to have declared the emergency too early than to have done it too late. If all turns out well, the likelihood of any penalty is rather remote, and its severity certainly negligible compared to the outcome in this case.—W. L. T.

What was the problem? It wasn't even very bad weather, though the visibilities were very low. But the controllers were uncooperative, and the pilot evidently became so fixated upon the idea of getting to an airport that he allowed at least one engine to run out of gas and then didn't restart it. Of course, if he was a VFR-only pilot in actual IFR conditions, the decision not to attempt a restart may have been wise, but in all likelihood a restart, with a prop still windmilling, would have involved no more than switching tanks. After a string of IFR-qualified pilots getting in over their heads in icing conditions we now see an equally deadly and even more unnecessary combination: a non-IFR pilot and IFR weather, however benign. The VFR pilot flies more or less wittingly into cloud, and then he must extricate himself. Since he has done something illegal, he is reluctant to put himself at the mercies of the controller. The controller, on the other hand, knowing that VFR pilots flirt with clouds every day, doesn't want to generate an emergency himself; so both sides may wait until it's too late.

Sometimes they don't wait, and it's already too late.—P. G.

8.

UP TO HERE IN ALLIGATORS

A telephone weather briefing warned the private pilot of 35T of snow showers in all quadrants except west at Salt Lake City, his point of departure. Neighboring mountains and passes were forecast to be obscured in clouds and snow showers; moderate turbulence was forecast below 15,000 feet, and pilots had reported heavy icing at 11,000 feet to the north and south of Salt Lake City. The current weather at Salt Lake was 1,500 scattered, 5,000 broken and 8,000 overcast, with 20 miles visibility. Weather at the proposed destination, Twin Falls, some 150 miles to the northwest, was reported to be 10,000 scattered with 30 miles visibility. The pilot departed Salt Lake VFR at 0837. Radar service was provided by Salt Lake City until the flight was handed off to Hill Departure Control. The following transcript covers eight minutes of the flight (HIF = Hill Air Force Base Approach Control; SLC = Salt Lake City Approach Control; *indicates direct land-line contact between ground stations).

HIF—Three-Five Tango, radar contact, say your altitude.
35T—Ten thousand, five hundred.
HIF—Roger, 10,500.
35T—Hill Approach, 35Tango, over.
HIF—Three-Five Tango, go ahead.
35T—I am, ah, ah, VFR and I've got myself into some clouds and I am lost, ah, have you got me on radar?
HIF—Affirmative, 35Tango, and, ah, you're five miles west-northwest of the Salt Lake Vortac. What, ah, would you like to continue on, or will you see any VFR around you?
35T—I'm kinda in a little pocket here, ah, I can see here, but I'm surrounded by clouds completely.
HIF—Ah, roger, ah, roger, would you like to return to Salt Lake?
35T—Yes, I'd better.
HIF—Okay, stand by—we'll hand you off.
*SLC—What's 35Tango going to do, Hill?
*HIF—He just called and said he went into the clouds, and he's lost. We still have him, we're talking with him to see what he wants to do.
*HIF—Okay, 35Tango wants to come back. Do you still have radar on him?
*SLC—Yes, still do.
*HIF—Okay, you want him 126.8?
*SLC—That's fine.

*HIF—He's on a heading of 110 . . . coming at ya.

35T—Salt Lake Approach, 35 Tango.

35T—Three-Five Tango. Salt Lake Approach, seven miles northwest Salt Lake Airport, ah, what are your flight conditions?

35T—Ah, I'm completely in clouds, I'm VFR, and, ah, I need to get back.

SLC—Roger, what is your altitude?

35T—Eleven thousand feet and I'm heading 110.

SLC—Three-Five Tango, roger, descend and maintain 6,000, heading 110.

35T—Drop to 6,000 and, ah, main . . . and a heading of what?

SLC—Heading 110, and are you instrument-qualified?

35T—No, I'm not.

SLC—Okay, sir, maintain a steady heading of 110, descend to 6,000, and, ah, what, do you have VOR equipment on board?

35T—Yes, I do.

SLC—Okay, sir, if no transmissions received for one minute during this vector, proceed direct to the Salt Lake Vortac and hold west at 6,000, right turns, over.

SLC—Three-Five Tango, turn right, a slow right turn, heading 160.

35T—Right turn, 160.

SLC—And descend to 6,000, and what is your altitude now?

35T—Altitude is 11,000 feet.

SLC—Roger, have you started your descent yet?

35T—Yes, I have, but I just (unintelligible) gone past 160.

SLC—Roger, fly heading 160.

35T—(Unintelligible), I have to level off.

SLC—Okay, sir, level off in straight flight. Let me know when you are leveled off and flying straight . . . 35 Tango, it appears you're circling to the right. You are in right turns.

35T—I'm completely disoriented. I'm (unintelligible).

SLC—Three-Five Tango, you're in circle, you're circling, right turns, you're in right turns, it appears . . . center the needle. Center the needle, 35 Tango.

35T—Control.

SLC—Say again.

35T—I've lost control.

SLC—Okay, sir, you'll break out of the clouds shortly, the, ah, you're three miles west of the airport; you'll break out of the clouds at about 8,000.

SLC—Kick left rudder, kick left rudder, 35 Tango . . . okay, and watch your airspeed. Left rudder and watch your airspeed.

35T—I'm falling.

SLC—Say again, 35 Tango.

35T—Falling.

SLC—Say again . . . 35 Tango, you're three miles, four miles northwest of the airport; you should break out below the clouds.

SLC—Three-Five Tango, if you can level off and fly southbound, fly southbound on a heading of 160, if you can do it.

SLC—Three-Five Tango, Salt Lake Approach, do you hear?

SLC—Three-Five Tango, Salt Lake Approach, do you hear?

The radar target of 35T disappeared during these last transmissions only one minute after the pilot reported losing control. The airplane crashed into a salt-water marsh in a 40 degree nose-down attitude. Damage indicated that the plane had been in a slight bank to the right prior to impact. The wreckage burned, but all aircraft components were found at the point of impact, and normal flight-control continuity was established. There was no evidence of an in-flight fire. The flaps were up, and the landing gear was down.

This accident is unusual because the airplane remained intact and because there was plenty of altitude underneath the ceiling for the pilot to recover from his predicament—if the pilot was still functioning and if he knew how to recover. The plane probably did not crash in a spiral dive, since recovery from a spiral is logical and would instinctively occur to any pilot when the horizon became visible. The condition of the wreckage also strongly suggested that the airplane did not hit the ground in a high-speed spiral. It was all in one place and not badly shattered, possibly indicating ground contact in a spin. The spin seems even more likely because the landing gear was extended at impact. The airplane had an automatic system that extends the gear at about 90 knots when the throttle is closed or at about 72 when the throttle is open. It is true that the pilot could have extended the gear, but it seems more likely that the airspeed was low, activating the automatic system. A spin could explain why the pilot did not recover after breaking out of the clouds. Private pilots are not required to demonstrate familiarity with spins and the recovery from a spin is not logical, so it probably would not be instinctive. To the pilot of 35T, the spin would have been the second problem of the flight—one that he simply could not solve.

His first mistake was not adhering to the basic precept of VFR flight—remaining clear of clouds. Once in the clouds the pilot did the correct thing, telling the controller of his predicament. The controller's efforts to help were commendable, but some of the terminology used raises a question. The controller told the pilot to "center the needle," exactly what the pilot needed to do if the needle in question was the turn needle. The pilot could have been confused, though, if 35T was equipped with a turn coordinator, which does not use a needle to indicate rate of turn. Many relatively new pilots have probably never flown an airplane with a turn needle. This is no criticism of the controller, who was working with the pilot under difficult conditions. A life preserver with *Titanic* spelled incorrectly beats hell out of no life preserver any day. However, this does suggest that the FAA should develop a standard procedure to use whenever a non-instrument-rated pilot is caught in cloud.

One aircraft manufacturer takes a step in the right direction by offering a card that fits on the sun visor, outlining a procedure for VFR pilots who must descend through cloud. It reads:

1. Contact nearest FAA facility for ceiling in area to be penetrated.
2. Notify facility of intention to penetrate undercast.
3. When 1,000 feet or more above clouds, slow to 100 mph IAS; gear down, straight and level.
4. Adjust throttle and trim to produce a 100 mph, 500 fpm descent on the desired heading, hands off, while still in clear air.

5. Use rudder pressure only to maintain heading. Do not touch wheel or column through clouds.

This advice doesn't explicitly cover a pilot who has already entered clouds, but it could be developed to cover other situations.

During a VFR-in-cloud procedure, relaxing the pilot is of primary importance. Once in cloud, there should be no urgency to do anything but to keep peace with the instruments. No immediate attempts should be made to tune a VOR, turn to a new heading, or change altitude. The pilot should remain still, for just moving the head can induce spatial disorientation. Only after a few minutes should the pilot try to extricate himself from his predicament. First, gradually slow the airplane and trim it to fly at a speed near the center of the green area on the airspeed indicator. If it's a retractable, perhaps the gear should be extended. If a descent is part of the solution, it should be gradual—500 fpm is plenty. If a turn is necessary, it should be shallow—10 degrees of bank is enough. It's like finding yourself up to here in alligators. Try to run and those alligators will get you for sure. The only chance of success lies in remaining calm and tiptoeing to safety.—R. L. C.

To an outsider it may seem puzzling that, given the weather briefing, the pilot set out on the flight at all. The explanation is that even in stormy weather VFR flight through mountain passes is sometimes possible, and the only way to find out is to go up and look. How can you fly into a cloud without meaning to? Clouds that look clearly defined from a distance are indefinite and ragged close up. Sometimes a rainshower, through which VFR flight would be possible, blends into a cloud, through which it would not. Pilots learn about weather by struggling with it; this pilot died, as it were, in basic training. The omission from basic flight training until a few years ago of the elements of IFR flight and the omission still today of spin recovery training both played a part in sealing the pilot's fate, even though, in all particulars except that of straying into the weather in the first place, his actions were correct. The weather was good in Salt Lake City and good at his destination; he hoped for the best in between.—P. G.

9.

BUZZED

The pilot checked out a flying-club airplane for a full day's use and promised to return the machine by 7:30 the following morning. Shortly before three A.M. the pilot departed Queen City Airport, near Allentown, Pennsylvania, on a VFR return flight to his home base, Slatington Airport, which was 15 miles away. There is no record of a weather briefing. The Allentown weather was 400 overcast, two and a half miles visibility, light rain and fog. The pilot was not instrument-rated. The aircraft, which we will refer to as 91L, contacted Allentown Tower at 0255 EDT and requested help. The following transcript covers the conversation during the next 11 minutes (ABE = Allentown Tower; all times are Eastern Daylight).

TIME: 0255
91L—Allentown Tower, this is 91 Lima, I need some help, over.
ABE—Nine-One Lima, Allentown Tower: go ahead.
91L—Ah, I took off from Queen City Airport, and I didn't realize that conditions were as bad as they are, and I'm in a little bit of trouble up here. Do you know where I am?
ABE—Okay, 91 Lima, I'll give you a hand on that. The Allentown altimeter is 30.15. What is your present heading?
91L—I beg your pardon, I don't understand that present heading.
ABE—Okay, with reference to your directional indicator, what direction are you flying now?
91L—I'm flying about 24.
91L—Yeah, but I don't know where you are.
ABE—Okay, 91 Lima, ah, I'm just checking the scope right now, ah, would you please let me know what your altitude is, sir.
91L—I'm approximately 2,000 feet.

TIME: 0256
ABE—Nine-One Lima, I understand you are at 2,000 feet and, ah, your heading is southwestbound, 240.
91L—Approximately, yes.
ABE—Okay, 91 Lima, would you say your fuel on board, please.
91L—Ah, I've got about a full tank.
ABE—Okay, 91 Lima, just stand by, and I'll be able to help you out in just one minute. Just stand by a minute.

TIME: 0257

ABE—Ah, 91 Lima, Allentown: would you look at your directional indicator again, please, and, ah, tell me what your heading is now, ah, if it's ah, northbound.

91L—Yeah, I'm flying north now.

ABE—Okay, 91 Lima, I have you in radar contact, and you are only one mile from the Allentown Airport. It's off to your right side. I'll turn the approach lights up as high as they go. Ah, let me know if you have them in sight off to your right side.

TIME: 0258

91L—Thank you.

ABE—Okay, 91 Lima, did you make a right turn?

91L—Yes, I made a right turn.

ABE—Okay, very good, you're gonna go right over the Allentown Airport; you're almost over it at this time. I'll turn, ah, all the airport lights on, ah, as high as they go for ya.

91L—Thank you.

ABE—Ah, 91 Lima, can you see the ground at all?

91L—No, I can't see the ground at all. I'm at two, I'm at 2,300 feet now.

TIME: 0259

ABE—All right, 91 Lima, are you familiar with an ILS approach?

91L—I'm not familiar with an ILS approach. I'm only a VFR pilot.

ABE—All right, 91 Lima, will you be able to, ah, fly the heading that I assign? If I give you a heading, will you be able to fly it?

91L—I think so.

ABE—All right, very good, and ah, just keep an eye on your altitude and maintain 2,500 on the altitude, and say your present heading now, is, ah, are you, ah, are you southbound?

91L—I'm heading six now.

TIME: 0300

ABE—All right, 91 Lima, I want you to turn to the left to a heading to 270. I want you to fly straight westbound.

TIME: 0301

ABE—Okay, 91 Lima, Allentown: I just checked with Wilkes-Barre, and up there they have, ah, 1,700 overcast with six miles' visibility. Now, I can attempt to vector you to the Allentown Airport, or, if you like, you can navigate up there in VFR conditions. It's your choice, sir.

TIME: 0302

91L—Let me try Allentown one time, then I'll go to Wilkes-Barre if I have to.

ABE—All right, sir, fine. Now are you headed westbound at this time?

91L—Yes, we're headed west now, approximate west.

68

ABE—All right, 91 Lima, are you in VFR conditions at this time?

91L—No, we are not. As a matter of fact, a plane just buzzed me now.

ABE—I, ah, missed the, ah, last transmission. Would you say it again?

91L—I said no we are not, and a plane buzzed me just now.

ABE—An airplane just buzzed you?

91L—Yes, an airplane just buzzed me.

ABE—Ah, 91 Lima, did you see what type aircraft it was?

ABE—Nine-One Lima, Allentown.

ABE—Nine-One, Allentown.

There was no further radio or radar contact with 91L, and at 0305 the police were informed that an aircraft had crashed into a house. The aircraft was 91L. The time of the accident was 0303.

Night VFR flight can be demanding in marginal conditions: in this case the conditions were far below marginal, so the outcome was no surprise. Perhaps the view from the airport ramp suggested conditions good enough for the short 15-mile transfer from one airport to another. If so, the eye deceived. Weather information was readily available, but the pilot apparently chose to fly without it. It is very hazardous to fly on a cloudy day without obtaining information on the ceilings and visibilities; furthermore, it's positively lethal to fly at night without doing so.

The preflight mistakes are clear. Once aloft, the pilot acted wisely when he called the tower at nearby Allentown and confessed his predicament. By so doing he enabled the controller to keep his flight from menacing aircraft operating IFR. Contact with the controller also assured that the pilot would get whatever aid was possible from the man on the ground. The transcript strongly suggests that the aircraft was operating in the clouds. When it moved over the airport, approximately 2,000 feet above the surface, even the brightest lights were not visible to the pilot. It is also clear that the pilot was doing a relatively good job of keeping the aircraft upright. Nothing he said indicated trouble with aircraft control. His heading did vary occasionally, but some of that was probably intentional.

The pilot's decision to try an ILS at Allentown instead of going to Wilkes-Barre would have made any necessary instrument letdown much less critical. But the flight would have been across the Pocono Mountains. There was no way to do that VFR under the conditions that existed, and the pilot might have reasoned that an airport a few miles away is worth several airports 50 miles away. Once the decision to try Allentown was made, the pilot seemed to do a creditable job of flying the westerly heading suggested by the controller.

Then came the mystery of the other airplane and the pilot's report that he had been buzzed. It's a quickly solved mystery: there were some radio towers—with their flashing red obstruction lights—that were probably poking up into the clouds in the vicinity of the crash site. It seems logical to theorize that the pilot flew near one of these and construed the red glow from an obstruction light to be a light on another airplane.

When a pilot flies solely by reference to gauges, the balance is delicate: even trained pilots become spatially disoriented when diverted or visually confused by an unusual outside cue. The trained pilot knows what to do, though, and he concentrates on flying the airplane by reference to the instruments and not by reference to any feeling he might have. In the case of 91L the diversion of a light on the tower might well have been the final straw. The pilot had been able to manage up to that point, but he must have been unable to cope with attitude control after becoming spatially disoriented.—R. L. C.

The possible final straw, spatial disorientation—it is also popularly called "vertigo"—is an annoying bugbear of instrument flight. Although the inner ear's balance mechanism cannot tell up from down for long in an airplane without some sort of visual assistance, it can produce very persuasive illusions about the pilot's position and motions. Instrument pilots learn to ignore these sensations and to depend on the instruments alone. If the pilot of 91L turned his head to follow the light that he thought was another airplane, he certainly experienced vertigo—rapid head movements are prime sources of the devilish sensations. At that point, it is not surprising that he lost control of the airplane. Another VFR pilot, in the following incident, hoping for the best, flew into the same sort of conditions—night IFR—but waited too long to call for help.—P. G.

10.

TOO LITTLE, TOO LATE

At 0533 on the morning of September 4, 1969, the Saginaw, Michigan Flight Service Station received a telephone request from the pilot of Piper Cherokee N4898L for a weather briefing for a flight to Gary, Indiana. At 0829 the pilot called again, requesting the en-route weather to Michigan City, Indiana. Finally, at 1517 the same afternoon he made a third call, this one concerning an intended flight to East Troy, Wisconsin. Each time, the pilot was advised that variable low-to-medium ceilings existed throughout the area, with restricted visibility and moderate to heavy scattered rain showers.

At approximately 1800, Cherokee 98L departed the airport at East Tawas, Michigan. No flight plan was filed. Accompanying the pilot were his wife and their two young sons. No radio contact was made with the flight for more than two hours. An air-traffic-control specialist based at the Pellston, Michigan FSS made the following written statement concerning his communication with N4898L: "At 2004, N4898L was heard calling Manistee, Michigan VOR on 123.6. After a second call was heard, I called the aircraft on 123.6 and advised the pilot that Manistee VOR was a state-operated facility without voice communication. I further advised if you wish to contact Manistee, call them on unicom frequency 122.8, but at this hour there may be no one at the airport. The pilot acknowledged my transmission and made no further statement as to where he departed or what his destination may have been. I gave N4898L the Pellston altimeter setting."

At 2051, the pilot contacted the Marquette, Michigan FSS and requested the unicom frequency for Manitowoc, Wisconsin. Two minutes later, the pilot again inquired of the Marquette FSS if fuel was available at Manitowoc and gave his position as being over Bailey's Harbor, Wisconsin. Bailey's Harbor is approximately 65 statute miles northeast of the Green Bay, Wisconsin VOR. Marquette advised the pilot to contact the Green Bay FSS for this information. The pilot was heard calling Green Bay, but no contact was established. The whereabouts of N4898L was then unknown until 2315, when contact was established on the emergency frequency, 121.5, with the Truax Field control tower at Madison, Wisconsin. The following is a transcript of these and subsequent transmissions (98L = Piper Cherokee N4898L; MDS = Madison Tower and Approach Control; MKE = Milwaukee Tower and Approach Control).

TIME: 2315
98L—Can anybody read me? This is 4898 Lima.

98L—This is 4898 Lima. Can anybody read me?

MDS—Nan 4898 Lima, Madison Tower reads you, go ahead.

98L—What is the weather condition at Milwaukee?

MDS—Stand by one, I'll check it.

TIME: 2317

MDS—Nan 48, Nan 4898 Lima, Milwaukee weather measured ceiling 1,000 broken, 2,300 overcast, visibility two miles, rainshower, haze, and smoke.

98L—What is their tower frequency?

MDS—Mitchell or Timmerman?

98L—Pardon me?

MDS—Milwaukee Mitchell or Timmerman?

98L—Timmerman Field.

MDS—Stand by.

MDS—Timmerman Tower 120.5, 120.5.

98L—Should there be somebody there now?

MDS—Stand by one, and I'll check.

TIME: 2318

MDS—Niner-Eight Lima, Madison Tower. Timmerman Tower closes at 10 o'clock at night. It's now 20 after 11.

98L—Nine-Eight Lima. Can you give me a frequency that is open there at Mitchell?

TIME: 2322

MDS—Nan 98 Lima, Madison Tower.

98L—This is 98L. I can't read you.

MDS—General Mitchell Tower frequency 119.1, 119.1.

98L—Have any, have any dangerous storms on a true of, uh, 120?

MDS—Madison does not have the area weather; it's strictly a tower facility.

98L—Are we on the emergency frequency?

MDS—Affirmative.

98L—I need help.

MDS—Do you know your present position?

98L—Heading 150 degrees, my omni shows—just a moment—shows to Milwaukee 130.

MDS—Stand by.

TIME: 2324

98L—Can you read me? This is 98 Lima.

MDS—Four-Eight-Nine-Eight Lima, Madison Approach Control reads you, go ahead.

98L—Please repeat.

MDS—Niner-Eight Lima, do you read Madison Approach Control? If you read, just key your mike for 15 seconds; key mike and no voice.

MDS—Just hold your mike keyed, 98 Lima; keep your mike keyed for 15 seconds.

72

MDS—Nan 98 Lima, you say you're on the 130 heading inbound to the Milwaukee VOR. Is that affirmative?
98L—Just a moment, please.

TIME: 2325
MDS—Nine-Eight Lima, did you say you're heading 130 degrees to the Milwaukee VOR? Is that correct?
98L—Nine-Eight Lima. That's affirmative. I'm in a heavy storm.
MDS—Stand by one. Are you able to maintain VFR conditions?
98L—This is 98 Lima. I'm in bad turbulence.
MDS—Nine-Eight Lima, this is Madison. Your present position—stand by one.
MDS—Nine-Eight Lima, are you familiar with Milwaukee area?
98L—Negative.
MDS—Okay, your present position is about 20 miles northwest of the Milwaukee VOR; you're about 10 miles west of the Hartford Airport.
98L—Where do I go?
MDS—Are you able to maintain—are you inbound to Milwaukee?
98L—Give me the closest field. I'm low on gas.
MDS—Okay, key your mike again for five seconds. Just key it for five seconds.
MDS—Okay, if you take a heading, 98 Lima, if you take up a heading of approximately 090 degrees from your present position, you're about 10 miles due west of the Hartford Airport, and that's just north-northwest of the Milwaukee VOR on about the 340-degree radial; they got a lighted 3,000-foot black-topped runway at the Hartford Airport.
98L—Was the heading 190?
MDS—Zero-Niner-zero heading from your position; 090 heading, which would bring you right over the Hartford Airport.
98L—Nine-Eight Lima.
98L—I'm heading straight east.
MDS—Nine-Eight Lima, roger.
MDS—And 98 Lima, are you able to maintain VFR conditions?
98L—This is 98 Lima. Do you—still have me?
MDS—Nan 98 Lima, say your altitude.
98L—Twenty-Five hundred.
MDS—Okay, continue.
98L—I'm having a hard time holding course.

TIME: 2331
MDS—Nan 98 Lima, Madison Approach Control, over.
98L—Nine-Eight Lima.
MDS—Nan 98 Lima, what radial of the Milwaukee VOR are you crossing now?
98L—One one six point four.
MDS—What's your—what bearing from the Milwaukee VOR are you—what bearing from VOR?
98L—I didn't read your last transmission.

MDS—What bearing have you from the Milwaukee VOR? What's your bearing to the Milwaukee VOR?

98L—One-thirty.

98L—One-thirty-two reading.

98L—Can you read me?

98L—Can you read me?

TIME: 2333

98L—This is 98 Lima. Can you read me?

MKE—Ninety-Eight Lima, this is Milwaukee Approach Control, how do you read?

98L—Yes, give me a bearing, please.

MKE—Transmit five seconds, please.

MKE—Ninety-Eight Lima, fly a heading of 120.

98L—Which way do I go?

MKE—The closest turn to a heading of 120.

98L—How close?

MKE—Ninety-Eight Lima, just fly a heading of 120. I will give you a DF steer toward Mitchell Field as soon as feasible; I will try to pick you up on the radar.

TIME: 2335

MKE—Ninety-Eight Lima, did you copy "fly a heading of 120?"

98L—I'm having a hard time getting it; I'm not (unintelligible) approximately 30.

MKE—Ninety-Eight Lima, what is your type aircraft?

98L—PA-28.

MKE—Roger.

TIME: 2336

98L—I'm at 3,000 feet.

MKE—Roger, 98 Lima, are you flying a heading of 120 now?

98L—I can only go by the compass reading. I can't tell.

MKE—Roger, and how much fuel do you have on board?

98L—Very little; she's showing almost empty.

MKE—Roger.

98L—I think I'm on a heading of 140 now.

MKE—Okay, that'll be fine; then fly the heading of 140.

98L—I have a heading of 140; is that what you want?

MKE—That will be satisfactory for awhile—140. I read you loud and clear; transmit another five seconds, please.

98L—[radio keyed, voice in background] Just hold this.

MKE—Continue on your present heading.

98L—How far out are we?

MKE—Ninety-Eight Lima, we have no way to tell at the moment how far out you are. We are taking you in toward the airport and will get you toward the airport.

98L—Do you have any idea where I am?

MKE—We are attempting to plot you on our charts from DF steers from Madison and also from Milwaukee.

TIME: 2339
98L—Can I pick up another omni station?
98L—Can you give me another frequency for an omni station to work us?
MKE—Just remain on this frequency. We'll work you on this frequency, over.
98L—I'm almost out of fuel.
MKE—As close as we can plot from the DF steers from Madison and from us here, you're in the vicinity of around Beaver Dam area.
98L—What's my chances for any airport?
MKE—Do you have ground contact at present time, 98 Lima?
98L—Pardon me?

TIME: 2340
MKE—Do you have ground contact, or are you on top of the weather?
98L—I'm right in the weather.
98L—We're pointing on empty.
MKE—Ninety-Eight Lima, understand.
98L—We're at 3,000. Would we be better off lower?
MKE—Ninety-Eight Lima, I'm afraid if you go lower you may lose reception on your radio.
98L—Ninety-Eight Lima.
98L—Ninety-Eight Lima. We can't get a thing. We're at 4,700.
MKE—Okay, continue inbound. You're doing fine. You're coming toward the airport, and the course indicates you're coming right straight toward Mitchell Field, and when we pick you up on radar, we'll point out a closer airport, vector you to an airport as close as we can.
98L—I'll go in at Milwaukee.

TIME: 2344
MKE—Measured ceiling three-two broken; stand by, I'll get you the latest weather.
MKE—Milwaukee weather is measured ceiling 1,000 broken, 2,300 overcast, visibility two, light rainshowers, haze, and smoke.
98L—We can see a string of lights on the ground.
98L—They're just spotty, though.
MKE—Roger, 98L.
98L—I must be awful close to them; radio's changing awful fast.
MKE—Ninety-Eight Lima, roger. Transmit for about three seconds now.
MKE—Continue your present heading. It's taking you toward the airport.
98L—Have you any idea how far yet?
MKE—We estimate you to be about 15 miles from the Waukesha Airport, and this makes you about 27 northwest of Mitchell.
98L—Is that first one lit?
98L—Is that first field lit?

MKE—Waukesha does have lights, that's affirmative.
98L—You say it does or doesn't?

TIME: 2347
98L—Out of gas.
98L—What do I do now? The right tank just ran out of gas. We're on the left tank. Can you read me?
MKE—Roger.
98L—What'll I do?
MKE—Do you see any roads or light that you can land on?
98L—Negative, just occasional rain.

TIME: 2348
MKE—Did you make a turn or go northbound?
98L—Affirmative.
98L—What'll I do?
MKE—Are you still able to fly the aircraft?
98L—We're flying right now.
MKE—Do you still have fuel to proceed farther?
98L—I'm on my right tank; my other one went out.
MKE—All right, continue back on your one—all the way around on your 120 heading.
98L—I'm on it right now. What do I do?

TIME: 2349
MKE—Fly the heading of 120. We'll take you to the Waukesha Airport, which is about 12 miles away. Continue on your present heading.
MKE—Are you in a position to climb at the present time with your present fuel?
98L—Yes.
98L—Where am I now?
MKE—You are 10 miles away from the Waukesha airport.
98L—What about my direction?

TIME: 2350
MKE—Your direction is good. Just keep on the present course. Just keep your present course unless I advise something different.
98L—What's my course?
MKE—You are flying 130 heading, which will take you right into the airport.
MKE—Ninety-Eight Lima, what is your altitude now?
98L—Three-zero-zero.
MKE—Roger.
98L—Give me an altimeter reading.
MKE—Altimeter 29.92.

TIME: 2351
98L—We just conked out of gas. What'll I do now? We're out of gas.
MKE—Look for an open field, and we have you plotted on radar, so there'll be assistance there shortly.

76

98L—We'll take it down on the road.
MKE—That'll be good, 98 Lima.

TIME: 2352
MKE—Nan 98 Lima, when you get on the ground and as soon as feasible, call the Sheriff's Department at Waukesha, and we'll work from there.

There were no further radio contacts with Cherokee N4898L. An immediate search for the downed aircraft was initiated, and a short time later the wreckage was located approximately six miles from the Waukesha County Airport. A detailed examination of the crash site disclosed that the initial impact occurred on a large hickory tree at a point 35 feet above the ground. Approximately 65 feet east of this tree lay the separated right wing of 98L. The wing had made contact with the tree at a point four feet inboard of the tip. The impact was of such force that the leading edge was indented past the midpoint of the wing chord. About 150 feet past the wing, first ground contact was made, with the wreckage and debris line continuing another 45 feet. The fuselage came to rest in an inverted position. Fire did not occur. The pilot, his wife, and one son perished in the crash; the other son survived, however, with serious multiple injuries.

The sequence of events in the final flight of N4898L follows closely the script of many general-aviation weather accidents. The 29-year-old businessman-pilot had only 118 hours total time, 34 of which had been flown in the 90 days before the accident. Possessing only single-engine land rating, the pilot was hardly equipped to conduct a night flight in marginal weather. From information later obtained it was confirmed that the pilot's intended destination was East Troy, Wisconsin. The repeated weather briefings that he obtained during the day only confirmed the presence of poor weather along the route of flight. What prompted him to attempt the flight just before dark will never be known.

The first radio contacts with Pellston two hours after departure indicate that the pilot was near Manistee, a considerable distance north of his intended route. This was probably due to his attempts to find the best weather conditions. The next contact, with the Marquette FSS, confirmed that the Cherokee was over Bailey's Harbor and that the pilot had crossed Lake Michigan. For more than two hours, 98L wandered along its indefinite course before the pilot considered his position to be so desperate that he chose the emergency frequency to gain assistance. At this point, his situation had already become virtually hopeless.

The next 37 minutes record the desperate, fruitless effort by Madison and Milwaukee Approach Controls to find an available field for 98L. The state of mind of the pilot, made evident by his pleas for assistance, is only too clearly revealed in the transcript. The pressure of the moments after fuel exhaustion can only be imagined. The fact that the rate-of-climb indicator showed 500 feet per minute down, with the airspeed indicator reading 170 mph—both instruments frozen by impact force—gave mute testimony to the pilot's final loss of control over Cherokee N4898L. Had the flaps been extended and the airplane trimmed for an 80-mph glide the impact forces would have been reduced and the outcome perhaps rendered less tragic.—W. L. T.

It's one of the commonest situations—so easy to get into, so difficult to get out of. The next incident is almost an epitomization of this kind of predicament.—P. G.

11.

"AIRPORT IN SIGHT"

The guard at Wings Field, near Philadelphia, saw the single-engine aircraft take off at 3:35 A.M., but he didn't obtain its number because of darkness and fog. Just over 30 minutes later the pilot contacted Philadelphia Approach Control. The transcript begins at that time. We will refer to the fixed-gear single as 58L (PHL = Philadelphia Approach Control; PNE = Philadelphia Flight Service Station; 66 = an aircraft that relayed messages; * = conversations between third parties; times are Eastern Daylight).

TIME: 0406:58
58L—Tower, this is N58 Lima on 119.0 for vectors to North Philadelphia.
PHL—Who's calling Philadelphia?
58L—Five-Eight Lima for radar contact.
PHL—Traffic, is it 58 Lima?
58L—Five-Eight Lima. Do you read me?
PHL—Five-Eight Lima, how do you read?
58L—(garbled).
PHL—Five-Eight Lima, Philadelphia, how do you read?
58L—(garbled).

TIME: 0408
PHL—Yes, what is your request, sir?
58L—I want radar vectors to North Philadelphia.
PHL—Are you VFR?
58L—Affirmative.
PHL—Squawk 1210, 1210, I believe, ah, North Philadelphia is IFR.
58L—Ah, affirmative.
PHL—Five-Eight Lima, maintain VFR and squawk 1210.
58L—Squawk 1210.
PHL—November 58 Lima, all right, I say North Philadelphia is IFR. Did you file an IFR flight plan first?
58L—Negative.
PHL—Is pilot instrument-rated, aircraft instrument-equipped?
58L—Negative.
PHL—I say, the pilot instrument-rated and aircraft instrument-equipped?
58L—I don't understand you.
PHL—I say, is, ah, or are you an instrument pilot so you can make an instrument approach to North Philadelphia?

58L—Affirmative.

PHL—All right, say your altitude and heading.

58L—Five thousand five hundred and 210 degrees.

PHL—All right, go over to—do you want to make an approach to North Philadelphia?

58L—Affirmative.

PHL—All right, you'll have to file a flight plan, ah, go talk to Philadelphia Radio. File a flight plan and return to this frequency and, ah, maintain VFR, sir.

58L—Go to what frequency and file a flight plan?

PHL—Stand by just one moment, please.

PHL—November 58 Lima, it's 122.6, 122.6, for Philadelphia Radio.

58L—All right, sir.

(Note: the conversation between 58L and Philadelphia Radio was not transcribed.)

TIME: 0420:48

*PHL—Approach.

*PNE—Approach, reference to this 58 Lima that you had contact Philly. Philly Radio.

*PHL—Yeah.

*PNE—Okay, information is that he has a full tank, he's, he's VFR pilot, he's not instrument-rated, and he's in instrument conditions.

*PHL—He is already in instrument conditions?

*PNE—He's in instrument conditions, and I can't give him a steer, a DF steer into, ah, North Philadelphia. We're IFR. I can't give a DF steer at all in IFR conditions.

*PHL—Okay, give him back to me.

TIME: 0423:38

58L—Philadelphia, this is 58 Lima, over.

PHL—Five-Eight Lima, Philadelphia.

58L—Ah, 58 Lima. I want a DF steer to North Philadelphia.

PHL—Five-Eight Lima, roger; squawk ident on code 1210, 1210. Say heading and altitude.

58L—Six thousand, zero-nine-zero.

PHL—All right, and it's no use going to North Philadelphia. They're indefinite zero obscured with zero on visibility and understand you're not an instrument-rated pilot.

58L—Affirmative.

PHL—Well, you can't very well land at North Philadelphia with zero-zero weather.

58L—Ah, where can I go then? Ah, where is there an airport that I can land VFR?

PHL—Squawk 1210, 1210, and we're checking on a VFR airport, and how much fuel do you have aboard?

58L—I have a full tank.

TIME: 0424:45

PHL—Full tank. What is that in minutes and hours?

58L—I've got about four hours.

PHL—About four hours. All right, sir, and you say you're at 6,000 VFR. Reset squawk 1210—and you're in the vicinity of North Philadelphia?

58L—I was in the vicinity of North Philadelphia, but I've been flying around for about 10 minutes and heading 060 degrees.

PHL—Ah, I don't have you in radar contact, and I don't have any beacon targets out there at all. Reset the squawk to 1210.

58L—I'm squawking 1210.

PHL—All right, ah, the closest thing I can see will either be Reading or Baltimore; ah, the closest VFR, Reading, has 1,000 scattered with estimated 4,500, visibility four with haze, and, ah, you want to go on course that way?

58L—Ah, affirmative, I'll go on up to Reading.

PHL—All right, can you home in on anything and proceed direct to Reading?

58L—I did not get that transmission.

TIME: 0426:26

PHL—I say, can you navigate on your own toward Reading?

58L—Yeah, if I can find it on the map here.

PHL—All right, say your heading and altitude.

58L—Heading 030 degrees.

PHL—All right, is your transponder on? I don't see any beacon targets at all in the general area that you're talking about. Is your transponder on? If so, reset, squawk 1210.

58L—It's on but, ah, might not, ah, must not be working. You sound like you faded out.

PHL—All right, I would suggest that, going, making, ah, ah, 180, proceeding back to North Philadelphia, and maybe we can work from there. If you get too far out, I won't be able to read you.

[There followed a period of conversation about heading, altitude and radar contact. The last was established near North Philadelphia, and the pilot took up a heading for Reading, navigating on the Pottstown VOR. At one point the controller asked if the pilot had checked the weather before departure and if it was IFR at Wings Field when he left. The answers were negative and affirmative. The transcript is resumed with 58L in the vicinity of Wings Field, again en route to Reading.]

TIME: 0451:21

PHL—I say, you just passed over the Wings airport, via, any chance, do you have any ground contact?

58L—Negative, I don't have any ground contact.

PHL—Roger.

58L—Can you hear me now? I switched radios.

PHL—November 58 Lima, you're loud and clear, stand by.

58L—I see the ground right directly under me now.

PHL—Like I say, you just passed the Wings airport. Ah, I don't know whether it's IFR or VFR there, ah, you're just about a half mile north of the Wings beacon.

58L—Ah, roger, I see the ground there, ah, maybe I'll try turn around and descent. Maybe I can see it.

PHL—All right, if you can make it in VFR conditions. Are you going to make a left- or right-hand turn?

58L—I'll make a left-hand turn.

PHL—November 58 Lima, did you make a right turn?

58L—I made a left turn. Ah, I see the ground there—if I can find the airport, hold on, stand by.

PHL—All right, you're two, ah, two miles northwest of the airport.

TIME: 0454:53

58L—I'm heading left now so, ah, ah, should make another left turn?

PHL—Left turn, heading of about 130. However, maintain VFR at all times; 130 is a suggested heading.

58L—Roger.

PHL—All right 58, N58 Lima, the airport is off your right at two o'clock, just about a mile.

58L—Message again.

PHL—Right off your right wing now, right wing, two o'clock and one mile.

58L—I didn't get your transmission, come again.

PHL—I say, the airport, Wings airport, is off your right, two o'clock, one mile.

TIME: 0456:23

*PHL—Sixty-six would you relay, tell him Wings airport is off his right wing one mile?

*66—Who am I calling?

*PHL—November 58 Lima.

66—Fifty-Eight Lima, Wings Field is just off your right a mile.

*66—I don't hear him replying at all.

*PHL—Yeah, okay, would you listen, please; it's a VFR pilot caught on top.

*66—Okay, what frequency's on?

*PHL—All right, ah, stand by just a moment.

TIME: 0457:02

PHL—November 58 Lima, Philadelphia Approach, ah, do you still have ground contact?

58L—Negative, I just lost contact, and I'm climbing back up now.

PHL—Roger.

PHL—November 58 Lima, you say you don't have ground contact any longer?

58L—(garbled).

PHL—Did you say negative?

58L—Affirmative, 58 Lima. I have ground contact now. I'm at 1,300.

PHL—All right, and, ah, your heading's what, about 270?

58L—Affirmative.

PHL—All right, the Wings airport is off your left wing, just off your left wing and just about a mile and a quarter.
58L—Am I right over the airport?
PHL—It's, it's just slightly off your left.
66—It's off your left wing a mile and a quarter.
*PHL—Sixty-six, tell him again, it's off his left wing, right now, one mile.
66—Okay, Wings Field is off your left wing, one mile, right now.
PHL—November 58 Lima, off your left wing now, just about a half a mile.
58L—Roger.

TIME: 0458:58
PHL—I show it's right below, off your left wing.
66—Right below you, off your left wing, 58 Lima.
PHL—November 58 Lima, you passed it up. Did you see it, sir?
58L—Could I have that transmission again?
PHL—I say, you just went by the airport, did you see it?
58L—Negative.
PHL—All right, I would suggest making a left turn around and going back to Pottstown, and we'll try to work you into the Reading airport.
PHL—If you make a left turn around, you'll probably go right over the Wings airport again.
58L—I'm making a left turn around now. You say the airport should be off my left wing?
PHL—That is correct, off your left wing just about one mile.

TIME: 0500:25
PHL—All right, 58 Lima, the airport's at 12 o'clock and a quarter mile, do you see it?
PHL—November 58 Lima, you're right over the airport now.
58L—Over the airport now?
PHL—That is correct.
*PHL—Sixty-six, will you tell him he's right over the airport now?
66—Five-Eight Lima, you're right over the airport, right now.
PHL—November 58 Lima, Philadelphia, are you back on the frequency?
*PHL—Would you call him and see if he reads you? Five-Eight Lima.
66—Five-Eight Lima, do you read 66?
58L—I've got the airport in sight.
PHL—All right, give me a call on the telephone when you land, please.
*PHL—Sixty-Six will you tell him that?

TIME: 0502:10
66—Five-Eight Lima, they would like for you to call on the telephone when you land if you could, please.
58L—Roger.

There was no further transmission from the aircraft. The guard at Wings Field stated that an airplane made a low pass over the runway from the south at around five A.M. but was too high and disappeared to the north. The plane was found 8/10 mile north of the airport at 8:15 A.M. The aircraft was destroyed. Wreckage distribution was on a heading of 180 degrees, and there was no indication of preimpact failure or malfunction. The aircraft clocks stopped at 0510. The toxicological report did not reveal any condition that would have affected the pilot's ability to conduct the flight.

What can you say? The pilot acknowledged over the radio that he did not check the weather and that the Wings Field weather was IFR when he took off. Once aloft and in contact with Philadelphia, the pilot seemed to have no great trouble flying the airplane or acknowledging instructions. Things progressed rather smoothly, considering the circumstances, as radar contact was established and the airplane was headed toward Reading.

The desire to land at Wings, the point of departure, must have been strong when the pilot was told of its proximity, but the controller wisely suggested that the flight proceed on toward Reading when Wings wasn't sighted after a couple of passes. The pilot got a nibble on the next pass, though. He reported seeing the airport, and a statement by the guard suggests that he was able to find and fly over the runway. The pilot was probably maneuvering for another approach to the runway when he either lost control of the aircraft or just flew into the ground at rather high speed.

The lessons about checking weather and not launching VFR flights in IFR weather are obvious. A less obvious lesson relates to the foggy conditions that prevailed. Frequently it is possible to see airport lights from directly overhead, even though there's not the slightest possibility of landing either VFR or IFR. Relatively shallow fog can reduce the forward surface visibility to zero, even though the stars may be visible from the ground and the lights may be visible from the air.—R. L. C.

The voice on the radio provided moral support for what would in itself have been an ill-advised enterprise. Trying to drop into an airport barely showing through a hole in the clouds is tricky in daytime; even a highly experienced pilot would hesitate to do it at night. But the controller's voice saying, "Go ahead" makes all the difference. The controller was trying to help, and he had to assume that the pilot would not take undue risks. Sometimes a controller is neither trying to help nor to hinder but just doing his job and trying to avoid annoyances.—P. G.

12.

"WE'RE IFR—YOUR INTENTIONS?"

At 2115 on the night of August 3, 1969 Cherokee N5611W departed Pensacola, Florida on a VFR flight to Tallahassee, with a planned en-route stop at Panama City, Florida. Aboard the flight was the solo pilot, a 40-year-old career serviceman, who held a private-pilot certificate and had logged 362 hours. He was not instrument-rated. The pilot did not file a flight plan, and there was no record that he had received a weather briefing. The aircraft arrived at Panama City without reported incident and departed at 0015, en route to Tallahassee. At 0101, N5611W contacted the Tallahassee Control Tower. The following is a transcription of those recorded radio contacts (11W = Piper Cherokee N5611W; TLH = Tallahassee Tower and Approach Control; FSS = Tallahassee Flight Service Station; WBA = Weather Bureau Airport Station; 476 = National Airlines Flight 476; * indicates communications between ground stations on direct land line).

TIME: 0101

11W—Tallahassee Control Tower, Cherokee 5611 Whiskey, how do you read?

TLH—Cherokee 711 [sic] Whiskey, loud and clear, over.

11W—Cherokee 5611 Whiskey, I'm coming up on the VOR, ah, how about a straight-in approach?

TLH—Cherokee 11 Whiskey, roger, make straight-in Runway 18, wind 120 degrees at five. Report two miles final.

11W—One-One Whiskey, roger.

TLH—Cherokee 11 Whiskey, what's your position now, over?

TLH—Cherokee 5611 Whiskey, Tower.

11W—Ah, 11 Whiskey.

TLH—Roger, 11 Whiskey, if you read now, the field is IFR, measured ceiling 900 broken, 3,000 overcast, visibility 10 miles, over.

11W—One-One Whiskey.

TLH—And 11 Whiskey, request your intentions, over.

11W—Say again?

TLH—Roger, the field is IFR, request your intentions, over.

11W—I'm not reading you. You are broken up for some reason.

TLH—Roger. The field is IFR, with a measured ceiling 900 broken. Request your intentions, over.

TIME: 0105
11W—I'll go on over.
TLH—You'll be overflying, is that correct?
11W—Right.
TLH—Roger.
*FSS—Yes, sir.
*TLH—Have you got a flight plan on Cherokee 5611 Whiskey inbound?
*FSS—Um, haven't got a thing inbound, sure don't.
*TLH—Okay, then. He just called me for landing, then said he's going to overfly.

TIME: 0106
476—National 476, Tower.
TLH—National 476, Tallahassee.
476—We're going to Jacksonville if you'll put us on request.
TLH—All right, sir.
476—Ready to copy.
TLH—Cleared to the Jacksonville Airport via direct Greenville Victor 198, maintain 1,500, squawk 1100, Jacksonville Center frequency 132.8 after departure.
476—Jacksonville direct Greenville Victor 198, 1,500, alpha 1100, 32.8.
TLH—That is correct.

TIME: 0110
476—National 476, taxi.
TLH—National 476, taxi to Runway 18 or Runway 36, wind 130 degrees at eight. Altimeter 29.93.
476—Okay.

TIME: 0112
476—We're ready when we reach the end.
TLH—National 476, roger, turn right on course, cleared for takeoff.
476—Okay.
11W—Tallahassee Control Tower, Cherokee 5611 Whiskey, how do you read me?
TLH—Cherokee 11 Whiskey, roger, make a right turn, fly westbound. Advise you that the field is IFR and traffic departing Runway 36, over.
TLH—Four-Seven-Six, if you read, hold in position.
476—Okay.
TLH—Cherokee 11 Whiskey, Tallahassee, do you read?
11W—I read you very well, 11 Whiskey.
TLH—Roger, the field is IFR. Make a right turn, proceed westbound, traffic departing Runway 36 will be going eastbound.
476—We'll just do a 180.
11W—One-One Whiskey.
TLH—Roger, 180 approved.

TLH—One-One Whiskey, advise when you're well clear of the, ah, approach course there for Runway 18. Traffic is waiting to depart, over.

TIME: 0113
11W—One-One Whiskey.
TLH—Cherokee 11 Whiskey, I do not have you in sight now. How far are you, over?
11W—Ah, 15 miles, over.
TLH—You say 15 miles?
11W—Affirmative.
TLH—Ah, roger, continue.

TIME: 0114
476—What's it look like now?
TLH—National 476, cleared for takeoff and the Cherokee says 15 miles out; however, I think he's still up there just northwest of us.
476—What altitude is he?
TLH—He appeared to be 900 feet; he's VFR.
TLH—Cherokee 11 Whiskey, what is your altitude, over?

TIME: 0115
11W—Twenty-five hundred feet.
TLH—Roger, 2,500.
*TLH—Go ahead, weather.
*WBA—Yeah, if you see any lightning, let me know. The radar indicates increasing activity all through the area here, and I don't see a cotton pick—ah, any lightning.
*TLH—No, I haven't seen any yet. I heard center talking to Gainesville a while ago, ah, and they said they had a big line down there.
*WBA—Yeah, there's some stuff, ah, 42,000 feet, down around Perry here, and I don't see it. I don't see any lightning here.
*TLH—Well, I hadn't actually been watching up to now, to tell the truth. I've been, uh, filling out this traffic and all and haven't paid much attention. If I see any, though, I'll let you know.
*WBA—Okay, fine, thank you.
TLH—National 476, contact center on 132.8.
476—Okay, 32.8, goodnight now.
TLH—Goodnight.
TLH—Cherokee 5611 Whiskey, Tallahassee Tower.

TIME: 0116
11W—Tallahassee Tower?
TLH—Roger, 11 Whiskey. Tallahassee weather is measured ceiling 900 broken, 3,800 overcast. The field is IFR and, ah, request your intentions, over.
11W—I'm going over to Perry, Florida.
TLH—Roger, you're proceeding to Perry, Florida, is that correct?
11W—Affirmative.

TIME: 0117
TLH—Roger.

No further contact was made with N5611W. At approximately 0200 four witnesses reported hearing the engine sounds of a low-flying aircraft about 17 miles east of Tallahassee Airport. All witnesses reported substantially the same facts, and the statement that follows is representative of their impressions: "The sound of the airplane woke me up. It was south of my house and the sound of the engine was loud and then died away, then was loud again. This kept on as the plane moved to the west of my house. Then the engine began the last loud roar, and it sounded as if it was wide open in a dive until I heard it hit the ground. It was cloudy and drizzling rain at the time."

At 1030 the following morning the wreckage of N5611W was located in a pasture. The pilot had died in the crash. The aircraft had struck the ground in a near-vertical attitude, and a piece of the pitot tube was found stuck in the ground at about a 75-degree angle. No evidence of inflight structural failure or flight-control malfunction was found. There was no fire.

The night cross-country flight of N5611W had proceeded normally with no adverse weather problems until approaching its destination airport. When the pilot reported over the VOR for a straight-in landing approach less than 10 miles from the airport, the successful completion of the flight was all but assured. However, three minutes later the pilot was advised that the field was suddenly IFR with a ceiling of 900 feet broken, although the visibility remained excellent at 10 miles. At this time, N5611W was established on final approach, and no doubt the pilot had the runway clearly in sight. He now experienced some minor difficulties with his radio and may not have completely understood all transmissions from the tower. He was obviously confused as to the proper procedure to follow when asked his intentions.

Although it is true that the airport was technically IFR, N5611W was certainly not IFR, for he circled near the field at 900 feet and was observed by tower personnel. Since the aircraft was established on final approach well within the control zone when advised of the field weather conditions, a suggestion by the tower that the pilot continue the approach under special-VFR conditions would have been a reasonable solution to the problem. The inexperienced pilot may not have been aware of this. Instead, the pilot was effectively refused landing clearance and pressed to clear the area as rapidly as possible to allow for the departure of an airline flight. During the period when N5611W was turning away from the airport and the airliner was taking off, the weather bureau contacted the tower. Mentioned in this conversation were thunderstorm cells that rose to an altitude of 42,000 feet in the vicinity of Perry.

Less than two minutes later, N5611W was again asked his intentions. The pilot advised he was continuing to Perry. Although the tower operator was aware of the violent weather in the Perry area, he made no mention of it to the Cherokee pilot. The final transmission from the tower to N5611W came at 0117, a simple "Roger." The pilot was dead 43 minutes later.

The official probable cause listed for this accident reads: ". . . improper inflight decisions or planning, attempted operation beyond experience and ability level, continued VFR flight into adverse weather conditions, resulting in spatial disorientation."

While there is little doubt that the pilot did not take advantage of the many services available to him, such as weather information, flight plan, and advice and assistance from the tower, that was accessible had he made a simple request, his experience level no doubt led him to accept directions from the tower without question or evaluation. Although the tower operator on duty at Tallahassee performed his obligation to the letter of the regulations, one cannot help but guess on the final outcome of this flight had he extended himself to assist a private pilot with little weather experience on a stormy night in August. —W. L. T.

A piquant case of the tyrrany of the controller's voice. As pilots gain in experience, they gain a certain cynicism also, and, like the older students in any school, they learn to beat the system. They'll argue with controllers, refuse clearances, or give false position or altitude reports to keep the controller happy. In this case, they would, in any event, have continued the approach, IFR be damned. In the worst case you can always claim that you didn't hear the transmission. The controller's voice can be anything— parent, friend, enemy, God, seducer, betrayer, supporter. But the controller is only human.—P. G.

13.

"PLEASE KEEP TALKING TO ME . . ."

Early on the morning of June 6, 1969 a 24-year-old woman arrived at an FBO's office on the Opa Locka, Florida Airport. She had flown this operator's aircraft for about seven months and had been a private pilot since February of the previous year. On this day, she said that she intended to fly to Washington, D.C. and scheduled an aircraft—a Cessna 172, N8040L—for a three-day trip. The pilot was alone when she took off at 0700.

At 0800 on the same day, the Miami Weather Bureau Office received a telephone call from a woman who requested a weather briefing for a flight from Hollywood, Florida to Nassau in the Bahamas and then to Kingston, Jamaica. Shortly after this, a VFR flight plan was filed for Cessna N8040L, showing an intended flight from North Perry Airport, in Hollywood to Montego Bay, Jamaica. Intermediate stops were planned at Nassau, Georgetown and South Caicos Island. The flight plan also indicated an estimated speed of 100 knots, with four hours of fuel aboard and a total of 11 hours estimated time en route. Listed as being aboard was the pilot and one passenger. The estimated time of arrival at Montego Bay was 2000, with Kingston, Jamaica as the alternate airport. The remarks stated that a life raft and life jackets were aboard.

The flight plan was activated at 0900. The aircraft arrived in Nassau at 1104 and departed at 1315. The exact time of arrival at Georgetown, Exuma Island, is unknown, but, records show that N8040L departed at about 1630 with four and a half hours of fuel aboard. The next contact with the flight came well after nightfall, approximately three hours later, when the control tower on Grand Turk Island, British West Indies, heard a radio call requesting assistance. The 1855 weather report from Grand Turk indicated scattered clouds at 2,000 feet, visibility seven miles, wind 100 degrees at 16 knots; there was no moon. A transcript of the recorded radio contacts follows (40L = Cessna N8040L; GTR = Grand Turk Radio; 26D = unidentified aircraft; GTT = Grand Turk Tower; SCI = South Caicos Island; 869 = Pan American Flight 869; UNK = unknown aircraft).

TIME: 1934
40L—. . . Read me (very weak signal)?
40L—I'm receiving you garbled, South Caicos, this is 8040L, do you read me (strong signal)?
GTR—Eight-Zero-Four-Zero Lima, this is Grand Turk Radio.
40L—Would you repeat, please, 40 Lima.
GTR—Four-Zero Lima, this is Grand Turk Radio.

TIME: 1935

40L—Grand Turk Radio, that's wonderful! Can you tell me how to get to your transmitter? I crossed two sections of land—and—two very large islands—and my heading now is 125. Do you know where I am?

GTR—The approximate shape—say again the approximate shape of the two large islands, please.

40L—One rather elongated and the other more round—or oval—or kinda diamond-shaped (signal begins to fade).

GTR—Zero Lima, Turk, stand by.

40L—. . . read you fiver; 40 Lima.

GTR—Four-Zero Lima, okay, you say your heading is 125 degrees and you see two large islands. Is that correct?

TIME: 1936

40L—Grand Turk, this is 8040 Lima; I have passed the two large islands. There is another island dead ahead of me at 120 degrees.

GTR—Roger, stand by one moment, please.

40L—Grand Turk, I passed the two islands that I spoke of, and there is an island straight ahead of me at 120 degrees (fadeout).

GTR—Four-Zero Lima, understand, do you see any villages or anything on the island?

TIME: 1937

GTR—Four-Zero Lima, Grand Turk Radio, do you copy?

40L—Grand Turk—40 Lima, I'm not . . . (fadeout).

UNK—Four-Zero Lima, do you read (breaks up), do you read (breaks up) Delta, over?

40L—I read you that time, 40L (weak).

26D—Grand Turk, are you reading 26D?

GTR—. . . is Grand Turk. You are barely readable, and 40L is almost out of my range, come on (unreadable transmission).

TIME: 1938

GTR—Two-Six Delta, I am just barely reading you, sir. Can you switch to 121.5?

40L—Four-Zero Lima (fadeout), 30 minutes' fuel (very weak).

26D—Four-Zero Lima, are you reading Grand Turk?

GTR—Two-Six Delta, Grand Turk, are you reading me?

26D—. . . Grand Turk . . . 26 Delta, over.

GTR—Two-Six Delta, Grand Turk, go ahead.

26D—Okay, did you read—30 minutes' fuel (fadeout) three islands in a line?

GTR—Two-Six Delta, switch to 121.5, please.

40L—Grand Turk (very, very weak)—Grand Turk, this is 40 Lima (stronger) —would you repeat, please?

TIME: 1940

GTR—Four-Zero Lima, switch to 121.5.

40L—Grand Turk, I'm sorry, would you repeat, please, one more t (fadeout).
GTR—Four-Zero Lima, pan, pan, pan, (heard on 121.5 very weak).

TIME: 1943
GTR—Four-Zero Lima, do you copy? This is Grand Turk on 121.5.
40L—(very, very weak) Four-Zero Lima, I did not read (fadeout)—island—
zero . . .
GTT—Four-Zero Lima, this is Grand Turk Tower, how do you hear me?
40L—Four-Zero Lima, I read you well.

TIME: 1953
GTT—Roger, give a short count, please.
40L—One, two, three, four, five—do you read me?
GTT—Four-Zero Lima, you are loud and clear. Say your present heading and airspeed.
40L—Grand Turk, this is 40 Lima. If you were transmitting to me just now, I missed you.
GTT—Four-Zero Lima, say your heading and airspeed, over.

TIME: 1954
40L—Grand Turk, this is 40 Lima, do you read me?
GTT—Roger, 40 Lima, reading you loud and clear. Say your heading and airspeed.
UNK—Four-Zero Lima, say your heading and airspeed.
40L—Grand Turk, I do not receive you well now; you are garbled (signal strong).
GTT—Aircraft transmitting, get her heading and airspeed, please.
UNK—Nine-Four-Three, say your heading and airspeed, go ahead.
GTT—Four-Zero Lima, turn on your landing lights and say your heading and airspeed. Do you read? Go ahead.
40L—Would you please repeat one more time? Four-Zero Lima.
GTT—Roger, 40 Lima, turn on your landing lights; say your present heading and airspeed, please.
40L—Four-Zero Lima. Landing lights are on, my airspeed is 100 knots, altitude 650 feet, my present heading is 180 degrees.
GTT—Roger, 40 Lima, how many persons aboard your airplane, over?

TIME: 1956
40L—This is 40 Lima. I see an airplane—straight ahead of us—a heading of 120.
GTT—Roger, Grand Turk transmitting in the blind to all aircraft in the Grand Turk area heading 120 degrees. I have a lost aircraft, 650 feet. She has her landing lights on, approximately 15 minutes fuel. Does anybody have her in sight?
40L—Grand Turk, this is 40 Lima, I made an error: it was not an aircraft; it was a star—I was looking too hard. Four-Zero Lima. I'm sorry.

TIME: 1957

40L—Grand Turk, this is 40 Lima, I will circle over this island till you can give me an approximate heading to Turk. I think I might be able to make whatever you tell me. At least I will try. I have survival equipment aboard and one male passenger. If you can tell me where I am, if you can locate me . . .

GTT—Four-Zero Lima, do you have any islands in sight?

UNK—Four-Zero Lima, do you have any islands in sight?

GTT—Aircraft transmitting, Turk, say your call sign.

869—Eight-Six-Niner.

GTT—Eight-Six-Niner, say your position.

869—Eight-Six-Niner, we're on Bravo 12.

GTT—Roger, can you pick her up on your bird dog?

869—We have negative VHF bird dog.

GTT—Roger, Turk.

TIME: 1959

40L—Grand Turk, 40 Lima, do you read me?

GTT—Hello, 40 Lima, reading you loud and clear. How me?

40L—There seem to be any way out of this situation?

GTT—Four-Zero Lima, do you have an island in sight, over?

40L—I have . . .

GTT—Do you have survival gear on board, over?

40L—. . . an island in sight. I am remaining over this circular one, because I do not know what else to do.

TIME: 2000

GTT—Four-Zero Lima, how long have you been flying from Georgetown to the island you're over now?

SCI—Four-Zero Lima, do you read South Caicos?

40L—This is 40 Lima, ah—I didn't receive you well that time, would you please repeat?

SCI—This is South Caicos. Now do you hear?

40L—I'm sorry, I read you garbled, please try once more.

GTT—South Caicos, maintain radio silence. Four-Zero Lima, understand that you are receiving me loud and clear. Is that correct?

40L—Yes, I am receiving you loud and clear. I sometimes have a little difficulty but most of the time am receiving you loud and clear, yes.

GTT—Four-Zero Lima, do you see any lights on the island?

40L—The two islands are totally deserted.

GTT—Four-Zero Lima, you're very, very close to Grand Turk. You are very close at 650 feet. I'm going to turn the runway lights up to maximum. Look around and see if you see them coming up.

40L—Four-Zero Lima, there are a lot of clouds in the area—it is difficult to s— to see.

GTT—Four-Zero Lima, we have the runway lights turned up and you must be very, very close. Do you have it in sight?

TIME: 2003

40L—No, I do not. Would you give me a heading to come in on? I don't know where I am; I have remained circling over these two islands because I did not know where I was or where to go from here; however, I do know that I am fairly close to you. Do you have any idea where I am or what heading to take from here?

 GTT—Four-Zero Lima, what time did you take off from Nassau? Tell me how long you've been flying, over.

40L—I took off from Nassau at 17—Zulu, but I had a very long wait in Georgetown; I refueled in Georgetown—topped the tanks there.

GTT— What time did you take off from Georgetown?

40L—Grand Turk, I could not tell you exactly; I think it was around four o'clock, or four-thirty.

TIME: 2004

GTT—Four-Zero Lima, understand. How many hours of fuel did you have on the airplane when you took off?

40L—I had four hours of fuel from Georgetown, right now my gauges—one reads empty and one reads a little less than one-fourth.

GTT—Four-Zero Lima, Turk; continue circling the island that you have in sight and stand by. We can't give you a heading till we know where you are.

40L—Four-Zero Lima.

TIME: 2006

40L—Grand Turk, I really think I'm beginning to get into trouble. Do you have any idea where I'm at?

GTT—Four-Zero Lima, we do not have you fixed. Turn left 270, fly for three minutes. I'll give you a call if I don't have you in sight. We'll turn you around and go back to the island you have in sight. Do you understand?

40L—I'm remaining over the island I have in sight; however, we do have our Mae Wests just in case.

GTT—Four-Zero Lima, pick up a heading of 270 for three minutes.

TIME: 2008

40L—Grand Turk, this is 40 Lima, do you read?

GTT—Four-Zero Lima, Grand Turk, how do you hear me now?

40L—I cannot hear you; would you please repeat?

GTT—Any aircraft in Grand Turk area, advise 40 Lima to turn back to a heading of 090.

40L—. . . 090, I'm, on my way, thank you.

GTT—Roger, turn back to 090; you're going away from us now.

40L—I am on a heading of 090. I have left the island. Please keep talking to me—I need it.

GTT—Four-Zero Lima, Turk's reading you loud and clear. You're coming back to Turk now, and keep flying the airplane. How much fuel do you have left?

TIME: 2010

40L—. . . 090, I am at an altitude of 900 feet, cruising at 110 mph; we're at 2,200 rpm.

GTT—What color is the airplane?

40L—Is there a chance that I will make it to you?

TIME: 2011

GTT—Roger, 40 Lima, you're coming toward Grand Turk, coming in loud and clear. Continue this heading. We do not have a fix on you, and I can't give you a good heading, but you're coming back toward the island.

40L—But you do have a fix on me. Is that correct?

GTT—We estimate you to be near North Caicos, a runway at South Caicos, and you should come up on that island in just a moment.

40L—Are there—Grand Turk, are there any lights on South Caicos, or on Caicos?

TIME: 2012

GTT—Roger, 40 Lima, affirmative. There are lights on South Caicos. Now turn to a heading of 120.

40L—Roger, this is 40 Lima, I am now on a heading of 120 at an altitude of 1,200 feet, airspeed 100 knots.

GTT—Roger, 40 Lima, you're doing real well. Continue looking for the island; the lights are lit up at Grand Turk; the runway is lit up. You should be able to see the island shortly. Continue flying the airplane and say the type aircraft, please.

40L—. . . make the island but not South Caicos. Are there any highways on which I can land?

TIME: 2013

GTT—Four-Zero Lima, negative, there are no large highways that you can land on on the island, but there are runways at South Caicos.

40L—Grand Turk, do you estimate that I will make it as far as South Caicos with my fuel?

GTT—Roger, 40 Lima, we estimate you can make South Caicos with your fuel. What type aircraft are you?

40L—Cessna 172.

GTT—Four-Zero Lima, Cessna 172, continue flying the airplane, and do you have your landing lights still on?

40L—Affirmative.

GTT—Four-Zero Lima, give a short count to Clipper 444 on 121.5.

TIME: 2015

40L—Grand Turk, 40 Lima, one, two, three, four, five, do you receive me?

GTT—Four-Zero Lima, Turk, reading you loud and clear. Break, break! Clipper 444, you reading Grand Turk?

40L—I'm sorry, I did not receive your last information, repeat.

GTT—Roger, 40 Lima, do you have the island in sight yet?

40L—Four-Zero Lima, I can see no lights, and cloud formations make it difficult to tell. I'm sorry, I do not have it in sight.

GTT—Say your present altitude.

40L—My present altitude is 1,000 feet.

GTT—Four-Zero Lima, set your altimeter 29.89.

40L—Grand Turk, were you transmitting to me just then? I did not receive; I heard 989, that's all.

GTT—Four-Zero Lima, we have an aircraft, a Clipper 444, who is trying to give you a fix on 121.5.

TIME: 2017

40L—This is 40 Lima, I understand that there is an aircraft that is trying to get a fix on me. Four-Zero Lima. Do you receive me?

GTT—Four-Zero Lima, you're transmitting on 126.2.

TIME: 2018

40L—This is—this is 40 Lima. I do not receive you well and I am now out of fuel—I think. It's on empty, and I haven't got much left. My heading is 120.

This was the last contact with N8040L. From June 7 through June 10 seven U.S. Coast Guard and three U.S. Navy aircraft searched a 22,000-square-mile area. All vessels operating in the Caicos and Bahamas area were requested to maintain a watch for the aircraft. To this date no trace of the aircraft or of its occupants has been found.

There are no sectional or world aeronautical charts available that extend beyond about 70 miles from the coast of Florida. USAF world and operational charts are available, but these are approximately three and a half by four and a half feet in size, and two are required to cover the route followed by N8040L. These charts show nondirectional radio beacons on Grand Turk Island, South Caicos, and San Salvador Islands, but no radio frequencies are shown. Larger-scale ONC charts are also available, but they are no different in detail. The true bearing from Georgetown to South Caicos measures 115 degrees, with slightly over two degrees west variation at Georgetown, increasing to well over four degrees at South Caicos. The distance from Georgetown to South Caicos measures 265 nautical miles. The direct-line distance to Grand Turk measures 285 nautical miles. The only Government-published charts of the area that show the radio beacon frequencies are the low-altitude en-route instrument charts for the Caribbean and South America. These charts carry the low-frequency beacon frequencies as well as the magnetic courses and distances for amber airways to various points, but none crosses Georgetown. These charts are available throughout the Miami area, but no record of a purchase by the pilot was found.

The only navigational system available to the Cessna pilot was by non-directional radio beacons. The aircraft was equipped with a Narco ADF and two VOR receivers. Records of the pilot's training revealed that she had not received instructions in the use of the ADF and was not considered qualified for this type of radio navigation. The male passenger, later identified, had no record of any aviation training or experience. The flight-plan route of N8040L passed over an area dotted with islands, although much of the flight was over open water. Because of the pilot's lack of training and experience it is doubtful that the operator would have permitted her to rent the aircraft had he known her intended destination.

When the pilot finally established radio contact with Grand Turk Island, she was already facing a desperate situation. With only 30 minutes of fuel remaining, she was completely lost, darkness was rapidly approaching, and her low altitude made radio contact difficult. At 2006, the aircraft was circling a small, unidentified island, and the pilot was obviously reluctant to leave this last contact with land. At this time, the controller at Grand Turk began giving the pilot headings to fly, although the exact position of the flight remained unknown. The radio contacts show some confusion between the pilot and the controller regarding the headings given and the actual heading of the aircraft.

The pilot was first advised to take a heading of 270 degrees. This course was never acknowledged; in fact, the pilot stated that she was continuing to circle the small island. Finally, a second heading of 090 degrees was given and repeated by the pilot. This vector was intended to return the aircraft to its point over the island, but this may actually have been the first course taken to depart the island: shortly after 2008, the pilot stated: ". . . heading of 090. I have left the island." At 2012 the pilot was advised to take a heading of 120 degrees, as her transmissions became stronger. The aircraft remained on this final course until fuel exhaustion about six minutes later.

Since VHF radio communications were possible with the aircraft flying as low as 650 feet, it is certain that N8040L was in fact very close to Grand Turk Air Force Base. At this point even a minimal knowledge of ADF navigation would have led the pilot to a safe landing via the strong beacon located at the base. Since the pilot did not have this ability and the controller did not think to instruct her in it, however quickly or informally, only one reasonable course of action remained to her: a forced landing on the island directly below her. This landing would probably have been successful if it had been attempted while fuel still remained to allow the pilot a choice of landing sites. Instead, command of the aircraft was relinquished to the controller on the ground. This action alone was the most serious of all the many errors committed by the pilot, because the responsibility for the safety of the aircraft and its occupants remains with the pilot regardless of the circumstances. When direction of the flight was turned over to the Grand Turk controller, the pilot joined him in the desperate hope that the airport could be found before the final minutes of fuel were consumed. Although it is difficult to understand why the controller issued specific headings without knowing the precise position of the aircraft, it is equally difficult to understand why the pilot turned away from the only known land and accepted a course that led to a forced landing at night in the open sea.—W. L. T.

The argument that the pilot made a serious mistake in relinquishing command to the controller is unfair to the pilot: the controller sounded as though he knew where she was, and under the circumstances the best course was probably to follow his instructions. No doubt the controller felt certain that his guess about her location was right, but he got carried away with his role of savior. With a low layer of scattered cloud, it might have been hard to see lights, even nearby, from 1,000 feet. She should have landed on one of the islands—in a meadow, on a beach, in the water near shore. Once again, the possibility of a precautionary crash landing is one that controllers don't seem to consider. Perhaps a firm rule should be made: lost aircraft with under 30 minutes' fuel remaining should give specific consideration to a controlled crash landing while power is still available. It's a policy that would produce some ironies from time to time but perhaps prevent some deaths.—P. G.

14.

COURSE OF CONFUSION

A large, low-pressure area was moving slowly but inexorably across the northern Great Lakes into New England that autumn day. Sometime after midnight, the massive low stalled over northern New York State. Extending southward from its center, a warm front curved like an archer's bow across western Pennsylvania, central Ohio, and into the Carolinas, continuing to creep toward the Atlantic coast. At 0500 on the following morning, the copilot of an air-taxi turboprop (which we'll refer to as 21L) prepared to begin his work aboard the day's first scheduled flight. Before leaving home he assured his wife that he would return between five and six P.M. At 0600, he left to pick up the pilot of the flight, and together they arrived at La Guardia Airport, New York at approximately 0630.

The turboprop departed La Guardia at 0730 on its scheduled air-taxi route; this would be the only on-time departure that the plane would make throughout this stormy day. As the pilots proceeded with their routine of flights, they were met with continually deteriorating weather and associated air-traffic delays. Shortly after dark, the front had moved to a position along a line from southeastern Pennsylvania to North Carolina and out into the Atlantic Ocean. Associated with the front were widespread areas of precipitation with low ceilings and decreasing visibility. At 1900, the flight was again at La Guardia Airport, and the copilot called his wife to say that he wouldn't be home until nine P.M. At nine, the copilot again called to tell her that, due to delays in their schedule, he would not be home until 11 P.M. Shortly after 2200, 21L departed Kennedy Airport en route to Wilkes-Barre-Scranton Airport, Pennsylvania.

By the time the flight arrived at Wilkes-Barre-Scranton, at 2242, the front had become quasistationary, and weather stations along the route were reporting ceilings of 500 scattered, 800 overcast, with visibility five miles in light rain and fog. Cloud layers extended to 20,000. Weather radar reported scattered thunderstorm activity, with cells along the frontal line moving eastward at 20 knots. The entire weather pattern was characterized by a shallow layer of unstable air below 6,000 feet msl, with abundant moisture. The freezing level was reported as 11,400 feet.

At 2253, 21L contacted Scranton Ground Control for taxi instructions and IFR clearance to Allentown, Pennsylvania. The flight number for this segment of the flight was 65. Aboard Flight 65, in addition to the two pilots, were six passengers. The following is a transcript of the recorded radio contacts with the flight (SGT = Wilkes-Barre-Scranton Ground Control and Tower; SDC =

Scranton Departure Control; AAC = Allentown Approach Control; 065 = 21L Flight 65).

TIME: 2253
065—Scranton Ground, 65, instruments to Allentown.
SGT—Six-Five, Runway 4, wind 090 degrees at 15, altimeter 30.03.
065—Sixty-Five.
SGT—Six-Five, you're cleared to the Allentown omni direct to maintain 4,000. After departure it'll be runway heading for vectoring and squawk 0100, over.
065—Okay, ah, cleared to the Allentown airport directed to maintain 4,000, runway heading for vectors and squawk, ah, 0400.
SGT—Departure will be 126.3; advise this frequency when ready, and that's 0100.
065—Okay, sir, 0100, 126.3.

TIME: 2255
065—And 65 is ready.
SGT—Six-Five, cleared for takeoff.
065—Roger.

TIME: 2257
SGT—Sixty-Five, contact Departure, Departure Control.
065—What's the frequency?
SGT—One-twenty-six-three.
065—Tower, Departure Control is what freq. . .?
SGT—Roger, contact Departure Control 126.3.

TIME: 2259
065—Departure, 65 is out of 3,000 for 4,000. We'll call you level at 4,000.
SDC—Okay, are you receiving Allentown suitable for navigation?
065—Roger, we're proceeding direct Allentown.
SDC—Okay, you're proceeding now direct Allentown, radar service terminated, and you can contact Allentown approach, 118.4, reaching 4,000.
065—Roger.
SDC—See yah.

TIME: 2300
065—Allentown Approach, 65 level at four.
AAC—Roger, 65, report 12 north of the VOR, altimeter 30.05.

TIME: 2301
065—Allentown Approach, 65, do you read?
AAC—Roger, 65, loud and clear, do you read us?
065—Yeah, read you fine now.
AAC—Roger, report 12 miles north of the VOR, altimeter 30.05.
065—Approach, ah, 65, what kind of approach are, ah, do you allow this time?

AAC—Well, I wanted to wait till I was sure we had good communications before we got into that. The weather, 500 scattered, measured 800 overcast, five miles light rain and fog. It will be whatever you want.

065—How do you read this transmitter?

AAC—I gotcha loud and clear now, but when you first called, I think you were too far beyond the hills there. We couldn't pick you up that good.

065—Okay, thank you.

AAC—Did you get the weather okay?

065—Ah, we got the weather, sir.

AAC—Okay, what type of approach do you want?

065—Ah, how about the VOR, ah, down to, ah, south of the airport?

AAC—All right, plan on it, call me 12 north.

065—(microphone click).

TIME: 2303

065—Allentown, 65, ah, DME's not working too good, ah, we got, we'll give you the 030 radial of the East Texas VOR.

AAC—How far out do you estimate you are?

065—Estimate Allentown, I'd say about, ah, six, seven, ah, minutes, make it six minutes.

AAC—Six-Five, cleared for the VOR approach, landing Runway 6, wind 070 degrees at 10, report the VOR inbound.

065—Thank you.

TIME: 2323

AAC—Six-Five, what time do you estimate the VOR inbound?

AAC—Sixty-Five, Allentown, do you read?

There was no further radio contact with Flight 65. The wreckage of the airplane was located on the north slope of Blue Mountain, approximately 11 miles north of the Allentown Airport. Blue Mountain is a rolling ridge with elevations varying between 1,500 and 1,600 feet. The ridge runs in a general east-west direction and is located five and one-half miles north of the Allentown VOR. The aircraft struck the slope at an altitude of 1,540 feet, approximately on the 360 degree radial of the VOR. This is the proper inbound course for the VOR approach to Allentown.

An investigation showed that the aircraft had first impacted against some trees just prior to striking the ground. Broken tree limbs and various aircraft components, including both outboard sections of the wings, portions of the horizontal-tail surfaces, and the right engine, were distributed for 370 feet along a path of 180 degrees magnetic. An intense postimpact fire so damaged the cockpit instruments that an accurate documentation of their readings was not obtainable. Only two instrument readings were available: the altimeter setting of 30.02 and the time at which the aircraft's clock had stopped, 2314. The landing gear was in the up-and-locked position, and the flaps were extended. There was no evidence of powerplant, airframe, or control failure or malfunction prior to the crash.

100

The pilot, copilot, and two passengers died in the crash; four passengers survived with serious injuries. The surviving passengers reported that there was no indication of the impending accident. According to their statements, the curtains separating the cockpit from the cabin area were pulled closed before takeoff from Scranton, and the passengers were unable to observe the crew's activity during the flight. The survivors agreed that the engines sounded normal, although there was a great deal of turbulence and the flight was extremely rough. Immediately before impact, they had the sensation of being suddenly squeezed very hard into their seats. They reported that the aircraft caught fire immediately and that there were several explosions. The passengers remained strapped in their seats after the fuselage came to rest, and the survivors exited the aircraft before the fire reached the cabin.

The National Transportation Safety Board determined that the probable cause of this accident was the pilot's failure to adhere to the approved approach procedures and further cited the probability that fatigue had affected both pilots, who had flown extensively that day, much of the time under IFR conditions. On November 17, 1971, John H. Reed, Chairman of the NTSB, forwarded the following Safety Recommendation to John H. Shaffer, then Administrator of the FAA, concerning this accident. After stating the date and location of the crash, he continued: "Preliminary investigation and a review of the VOR approach procedures for the Allentown area indicate that the VOR approach plates may have been a factor in this accident.

"The aircraft crashed 10 miles north of the airport and five miles north of the Allentown VOR at an altitude of 1,600 feet msl. In an attempt to ascertain why the aircraft was at that point, the VOR approach plates were reviewed closely. It was noted that 1,600 feet is the low station altitude for the Allentown VOR approach to Queen City Airport, utilizing the East Texas VOR. The VOR approach to the Bethlehem Easton Airport (the destination of Flight 65), using the Allentown VOR, has a low station altitude of 2,200 feet.

"Since both instrument approach plates are entitled VOR-1 and have the word Allentown twice in proximity thereto, other factors might have caused the pilots to select the wrong approach plate for the approach they were conducting. Thus, when the aircraft was cleared for the approach and erroneously descended to 1,600 feet, there was inadequate terrain clearance, and the accident was inevitable." Reed's letter continued with several specific recommendations for FAA action meant to eliminate the possibility of this type of accident reoccurring. On November 26, 1971 Administrator Shaffer sent Reed the official FAA response, which said in part:

"The conditions cited in your recommendations have been reviewed. Based on available information, we do not consider that the VOR approach plates were a contributing factor in this accident, nor do we consider that the recommended actions are appropriate.

"The Board has based their recommendations on the premise that the pilot may have selected the wrong VOR approach plate due to some similarity between the airport names, haste on the part of the crew, or poor lighting.

"Our review of the ATC recordings indicates that the pilot was fully aware of his position and the navigation facilities he was utilizing and that he was familiar with the destination airport. The pilot reported that he would utilize the East Texas VOR approach to the south of the airport. He was subsequently cleared for the VOR approach to land on Runway 6 with instructions to report at the VOR inbound. In order to make a VOR approach utilizing either approach plate, a pilot must proceed to the VOR by airways at MEA and execute a procedure turn, since radar is not available and a straight-in without a procedure turn is not authorized. If the pilot selected the wrong approach chart, he would be required to make a procedure turn on the 284 radial, and the crash site, which was approximately five miles north of the Allentown VOR, does not substantiate this theory.

"In view of the pilot's voice transmission and the location of the crash site, we do not believe that he utilized the incorrect chart, that he was planning his approach in haste, or that he was having any cockpit difficulties.

"In summary, this was a routine, scheduled air-taxi operation. The pilot flew regularly into the Allentown, Bethlehem-Easton Airport and was familiar with the route and airport environments; therefore, the information available to the FAA does not support the action recommended by the Board."

The NTSB's concern that the crew of Flight 65 may have referred to the wrong approach plate offers an interesting and entirely feasible explanation for the aircraft's extremely low altitude on the inbound course to the VOR. The FAA rebuttal, which points out the wide variance between the two inbound courses, also offers a reasonable point of evidence to the contrary. The Allentown, Bethlehem-Easton Airport is located on the north side of the tricity metropolitan complex of Allentown, Bethlehem, and Easton, Pennsylvania. The Allentown Queen City Airport is located in the southern edge of this metropolitan area. The Queen City Municipal Airport lies close to the Bethlehem-Easton Airport and has a VOR-1 approach plate. The Queen City VOR-1 approach utilizes the East Texas VOR. The ABE VOR-1 plate uses the Allentown VOR for approach. The minimum altitude over the VOR inbound for the Queen City approach is 1,600 feet, whereas the minimum low-station altitude for the ABE facility is 2,200 feet. Both of these plates were filed together in the same plate binder used by the crew of Flight 65.

The greatest single difference between the two plates lies in the inbound heading to the final approach fix. The heading to Queen City is 104 degrees inbound, while the ABE approach requires a course of 180 degrees after a procedure turn. Because of these facts and the location and heading of the aircraft at impact, it is unlikely that the Queen City approach was being conducted, at least in the initial stages of the procedure. But perhaps we should consider another possibility.

At the time the accident occurred, the flight crew had been on continuous duty for over 14 hours. During this time, they had been fighting weather most of the day. The crew had been required to make numerous IFR approaches and had flown a total of five hours and 30 minutes in actual IFR conditions. One indication of crew fatigue may have appeared during the departure from Scranton when the pilot asked the frequency of Departure Control only two minutes after he had been given that information and had acknowledged it in the readback. We can assume that the proper approach plate was selected to begin the approach to Allentown, as is confirmed by the heading and course of the plane. However, during the inbound leg to the VOR, the pilot may have referred to the wrong plate to confirm his altitude to cross the VOR. If this was the case, it would explain why the pilot was at an altitude of 1,600 rather than the specified 2,200 for this segment of the approach. An easy mistake to make on a dark, stormy approach in turbulence.

As stated in the NTSB document, the two approach plates are identified in large print at the top of the page by the word ALLENTOWN, PA, with the smaller printing beneath this heading giving the name of the individual airports. The NTSB specifically recommended that the phrase "CAUTION—VERIIFY PROPER APPROACH" be added to these plates and to any others with a marked similarity. It is doubtful that this change would have prevented this particular accident. But it might be wise to consider changing the format of approach plates to make the airport name the most prominent heading.

Finally, one must comment on Shaffer's reaction. His response seems to be mainly involved with discrediting the NTSB theory as to the cause. While the FAA's defense of the dead crew is a rare and welcome event, Shaffer offered no theory as to why the pilot had descended well below the MDA in an area with which he felt the pilot was too familiar to misread the approach plate. The argument that the pilot would have proceeded to the VOR by airway and at an altitude consistent with the MEA seems to ignore the fact that the flight was on a "direct" clearance to the VOR and was not following a specific airway clearance.—W. L. T.

In analyzing an anomalous accident such as this one it is perhaps not very sensible to ignore the possibility that anomalous conditions led up to it. If the approach plate was in a binder and not clipped to the control yoke—different pilots have different habits— then it is quite possible that the pilot consulted one plate at one stage and a different one at another, particularly if a ventilator was open, since the very thin paper of the appproach plates blows around easily. Accident investigators always try to reconstruct the shortest, most logical chain of events that might have led up to an accident: more indirect, circuitous paths are ignored, even though real life is full of them. Often the clear, direct chain of events, even if it fully explains the accident, may not be what actually happened.—P. G.

15.

"USE ANY APPROACH YOU LIKE . . ."

The 26-year-old pilot, a professional flight instructor, departed his home base in San Carlos, California, accompanied by his wife, one afternoon in the summer of 1970. The purpose of the flight was a vacation trip to visit relatives in Oregon. The pilot first landed at Red Bluff, California, where he filed a VFR flight plan listing Portland, Oregon as his destination. At 0054 the following morning Cessna N5535D contacted Portland Approach Control and received an IFR clearance for a surveillance radar approach, which was accomplished without incident.

The pilot telephoned the Portland FSS at 0140 and filed an IFR flight plan to Astoria, Oregon. While he was taxiing out for takeoff, the pilot received his clearance to Astoria via Victor 112 at 5,000 feet. After takeoff and departure, the flight was radar-vectored to Victor 112 by Portland Departure Control. The following is a transcript of the pertinent radio contacts with Cessna N5535D as it proceeded along Victor 112 toward Astoria (35D = Cessna N5535D; PDC = Portland Departure Control Radar; SEA = Seattle ARTCC; * indicates direct communication between ground stations).

TIME: 0202
35D—Portland Departure Control, 5535 Delta, Portland Vortac outbound, over.
PDC—Three-Five Delta, roger.

TIME: 0206
35D—Portland Departure, 35 Delta, over.
PDC—Three-Five Delta, go ahead.
35D— Roger, are you going to handle me all the way to Astoria?
PDC—Negative. We'll be terminating radar service here in another 15 miles; Seattle Center radar is out of service.
35D—I see, thanks a lot.

TIME: 0213
*PDC—Ah, Seattle Center, 35 Delta is coming up on Pittsburg Intersection, I'll terminate.
*SEA—Yeah, that's fine.

TIME: 0214
35D—Three-Five Delta, Pittsburg Intersection, estimating Astoria in 23 minutes.

PDC—Cessna 35 Delta, roger, radar service terminated at, ah, Pittsburg, contact Seattle Center 124.2, over.

35D—Changing, thank you.

35D—Seattle Center, Cessna 5535 Delta, Pittsburg, 5,000 feet, Astoria.

SEA—Cessna 35 Delta, roger, 5,000, by Pittsburg.

TIME: 0221

SEA—Cessna 5535 Delta, cleared for an approach to the Astoria Airport. Astoria altimeter is 30.09; report initial and arrival to Astoria Radio, over.

35D—Roger, can I use a circle-to-land Runway 7 approach?

SEA—You can use any approach you'd like into Astoria that you're capable of making this morning. And weather looks about four miles in light rain and fog.

35D—(unintelligible).

TIME: 0226

35D—Three-Five Delta is out of 4,000, descending to initial approach altitude.

SEA—Cessna 35 Delta, roger.

This was the last contact with 35D. When the aircraft did not arrive at its destination, a search was initiated. The following day, the wreckage of the Cessna was located approximately 10 nautical miles east of the airport directly on the centerline of Victor 112. The aircraft had struck the east side of 2,780-foot Elk Mountain about 150 feet below the crest. The aircraft had impacted the 55-degree slope while in a wing-level attitude with the nose slightly down. A detailed investigation revealed no failure or malfunction of the powerplant, controls, or airframe prior to the crash. The navigation and communication radios were set to the proper frequencies and operating normally at the time of the crash. There was no fire, but both occupants died in the accident.

The National Transportation Safety Board reported the probable cause of this accident as pilot error due to a "descent below minimum safe altitude while under IFR conditions." The report did not speculate on possible reasons for the mistake. However, a careful review of the sequence of events preceding the accident makes the actual nature of the pilot's error quite obvious yet deceptively subtle.

The pilot had ample ratings to perform IFR flight. His total flight time was 1,560 hours, including 107 in the Cessna that he was flying and 169 recorded within the previous 90 days. Of his total flight experience only 18.7 hours were conducted under actual IFR conditions, but 88.1 hours were under simulated instrument conditions. The aircraft was equipped with a full gyro panel, but had only one navigation receiver and one communication transceiver. The plane was not equipped with DME.

The route of the flight, Victor 112, proceeds direct from the Portland Vortac to the Astoria VOR. The minimum en-route altitude for this segment of the airway is 5,000 feet, and the distance between the stations is 59 nautical miles. The Pittsburg Intersection is located 21 nm from Astoria. Since Seattle Center radar was out of service, the last precise location of the flight was determined when the pilot reported over the Pittsburg Intersection at 0214, level at 5,000 feet, estimating arrival at Astoria in 23 minutes. All communications between the plane and the ground indicate that the flight was proceeding normally.

At 0221 Seattle Center issued an approach clearance to 35D for the Astoria Airport. Five minutes later, at 0226, the pilot reported leaving 4,000 feet in his descent to the initial approach altitude. This was the last contact with the flight. Less than four minutes later and only 10 miles from his destination the pilot slammed into the cloudy peak. Each action of the pilot was reported to and acknowledged by the controller—he was precisely on course, had no known problems with his aircraft or its equipment, and carried the needed charts and approach plates to successfully conclude the flight. What went wrong?

The pilot made only one crucial mistake, and that error resulted in his fatal descent. This error was reported to the Seattle Center controller, but perhaps the controller did not realize the importance of what the pilot had said.

At 0214, when 35D called and reported over the Pittsburg Intersection, he estimated arrival at Astoria in 23 minutes. Seven minutes later, at 0221, the controller issued the approach clearance to Astoria. At this time, using the pilot's own computation of his groundspeed, 99 knots, we can determine that the Cessna was still some 25 nm from Astoria. Five minutes after this, at 0226, the pilot reported that he was descending through 4,000 feet for the initial approach altitude. At this time, he was more than 16 nm from his destination. The message was acknowledged by center without comment, and the stage was set for the crash. It is probable that the pilot interpreted the approach clearance as authorization to begin a descent to the initial approach altitude of 1,500 feet. It is also possible that he was unsure of this procedure, prompting his report, "out of 4,000 descending to initial approach altitude," which was actually intended to confirm his descent as the proper procedure. The controller's response probably seemed to the pilot to imply approval of his action.

Actually, the final response of the pilot confirmed that he was already 1,000 feet below the MEA for Victor 112 and continuing his descent into the mountains. Descent below an airway MEA is not really an unusual procedure in many terminal approach areas, but one fact that the pilot evidently failed to notice is an often ignored if not totally forgotten bit of critical information printed on nearly every approach plate—the minimum safe altitude. This notation on the Astoria VOR-7 approach plate shows the MSA for the sector within which Victor 112 intersects the VOR as 4,300 feet. Descent below this altitude is prohibited before reaching the VOR.

In retrospect it is a simple task to point out the errors that contributed to this accident. It is even easier to show that the elimination of only one of several mistakes could have prevented this tragedy. By assuming that a clearance is always correct, the pilot began preparing for his approach when the charts and plate indicated another course of action. The fact that the radar at Seattle Center was out of service probably contributed greatly to the apathy of the controller and his relative haste in releasing the flight early for the approach. A more definitive and precise clearance could have completely changed the final outcome of the flight. Add only the phrase, "Cross Astoria VOR at 5,000," and all doubt as to the correct procedure is removed. One could fault the controller even further for his lack of vigilance in not being aware that the Cessna was nowhere near the VOR when the pilot reported out of 4,000 feet descending.

Clearly, the correct procedure in this instance after receiving the clearance was to do nothing. This is unusual. When most clearances are received, immediate action is required. Of course, the pilot should have continued to maintain 5,000 feet until reaching Astoria and then descended to 1,500 for the procedure turn and execute the approach. Weather at his destination was reported as 500 feet overcast, visibility four miles in rain and fog, wind 180 degrees at five—below minimums for the VOR-7 approach but not for the VOR-13, yet at any rate hardly more than a routine procedure. However, after a long, tiring flight, only minutes from the destination, on a dark night, in the clouds, the pilot stiff, fatigued, and maybe a little anxious to get on the ground where friends and relatives wait, errors even in standard procedures are dangerously easy to make.—W.L.T.

This accident was premonitory, because the pattern was to repeat itself years later in the widely publicized crash of TWA 514, a Boeing 727 on an instrument approach to Washington National, which descended prematurely and struck a mountain. In the case of 514, the Air Line Pilots Association placed the blame on ambiguous wording in the regulations. Surveys of pilots showed that many did not know the exact implications of the phrase "Cleared for the approach." The TWA pilot and the pilot of the 150 made exactly the same mistake, but the mistake of the lightplane pilot was dismissed as a mere blunder of his own, and no attempt was made to improve procedures. If lightplane pilots had a union, the furor over 514 might have been over 35D instead.—P.G.

16.

SNIFFING FOR ASPHALT

The light twin left Utica, New York at 2235 EST, bound for Albany, where the weather was down to 100 overcast and 1½ miles visibility in freezing rain and fog. The flight proceeded normally and was handed off to Albany Approach Control at 2352 EST, at which time the transcript begins (110 = the twin; APC = Albany Approach Control; TWR = Albany Tower).

TIME: 2352

110—Good evening, Albany Approach, 110 here.

APC—One-Ten, this is Albany Approach Control, squawk ident. Albany weather measured ceiling 100 overcast, visibility one-half mile, light freezing rain, and fog. Advise your intentions, please.

110—Ah, roger. We'd like to continue and, ah, ah, we have a package for you people there from, ah, Utica Tower.

APC—Yes, ah, advise your intentions, ah, as to the weather conditions, please.

110—We'd like to continue inbound and hope the visibility will pick up to three-quarters of a mile.

APC—One-Ten, roger. The wind, is, ah, right now 100 at, ah, three knots, altimeter 29.75.

110—Thank you. Has it just gone down recently or what?

APC—Okay. Ah, it went down about 20 minutes ago; the, yeah, the runway, 19, ah, visibility is, ah, well, hang on just a second. I'll check that.

APC—One-Ten, the runway, 19; visibility is, ah, seven-eighths of a mile.

110—Okay. All we need is, ah, three-quarters, so it sounds good.

APC—That's right. The runway conditions—ah, this, ah, right now, almost three hours old—deicer and sand have been applied over a glazed surface, which was previously reported as 70% bare, with a thin layer of hard-packed snow and scattered bare spots along the edges. The braking action has been reported as fair to poor by all types of aircraft.

110—Okay. Thank you very much.

APC—We have an area of, ah, considerable precipitation, and it is directly over Albany at this time, and it appears to be about 25 miles, ah, well, actually, about 40 miles in diameter, moving very slowly; it is raining here now.

110—Ah, thank ya. Understand ya have light, ah, freezing rain now.

APC—Ah, ya, that's the, ah, weather observation—light freezing rain.

110—Thank you.

APC—And, ah, additionally, 110, we're using Runway 19, by the way.

Additionally the, ah, locator at the outer marker and the glidepath are out of service. Expect radar vectors, ah, localizer Runway 19 approach only.

110—Okay, thank you. Is the visibility any better landing Runway 1?

APC—No, I would say not.

APC—One-Ten, fly heading 090, descend to and maintain 4,000.

110—Ah, roger. 090 down to four.

APC—One-Ten, the, ah, last braking action report I had, Runway 19 was approximately, ah, oh, two and a half hours ago, and it was, ah, fair to poor and, ah, more on the poor side by a Beechcraft Baron.

110—Okay. Thank you.

APC—Should have a new weather observation, ah, momentarily.

APC—One-Ten, ah, fly heading of 100 degrees and understand there is a south wind, ah, at altitudes, approximately 20 knots.

110—Ah, roger.

APC—'Cause it's got you headed about 060 right now.

110—Yeah, it's quite unbelievable up here.

APC—Right down here, it's, ah, well, right now I'm showing 060 at two.

110—Well, that's not too hard.

APC—One-Ten, the, ah, weatherly observation is the same, measured 100 overcast, one-half mile, light freezing rain and fog, the surface temperature now 32°, and the dew point 30.

110—Thank you. Is that one-half of a mile or seven-eighths of a mile?

APC—The prevailing visibility is one-half. Runway 19 visibility is seven-eighths of a mile.

110—Okay. Thank you.

110—And 110 is level at 4.

APC—One-Ten, roger. I'll have the lower, ah, about five miles (Sound of two microphone clicks).

APC—One-Ten, turn right heading 100, descend to and maintain 3,000.

110—Ah, roger.

APC—One-Ten, turn farther right 120.

110—Roger.

APC—One-Ten, descend to and maintain 2,000.

110—Roger. Out of three for two—110.

APC—One-Ten, your position is six from the outer marker. Turn right, heading 160, cleared the ILS—correction, that's the localizer Runway 19 approach. Runway 19 visibility now one mile.

110—Ah, roger. We're cleared the localizer for 19.

APC—Six miles for the outer marker, 110.

APC—One-Ten, contact the tower, 119.5.

110—Thank you.

110—Good evening, Albany Tower, 110.

TWR—One-Ten, report the outer marker, cleared to land. The wind is now 130 at three to five knots.

110—Ah, roger.

110—And, ah, 110 is, ah, marker inbound, ah, and you did say the glideslope and, ah, compass locator were out?

TWR—That is correct, 110. You're cleared to land; wind is light and variable right now, ah, about 250 at three knots.

110—Ah, thank you.

110—One-Ten here again. Ah, we had a good indication on your, ah, compass locator, and we did get a light from the outer marker, so, ah, I don't know what's out, out here.

TWR—Okay. Well, we'll try resetting the one, ah, they're working on—the glidepath. Eh, got a flag on the glidepath?

110—Ah, negative on that, but, ah, got a good, good swing.

TWR—Okay. Ah, the compass locator did reset.

110—Okay.

TWR—Compass locator, compass locator is back.

110—Thank you.

(Sound of aircraft crash-locator signal of very short duration.)

The main wreckage of the aircraft was found 3,000 feet short of the runway. The wreckage path extended approximately 2,000 feet and was in line with the localizer course. Examination disclosed no malfunction of any system or component, and no fatigue failures were noted at points of structural separation. The cockpit area was destroyed by fire, so it was not possible to document any instrument or avionics-equipment settings. However, there was no indication that the airplane was in trouble before it flew into the trees and then into the ground. The NTSB's report on this accident listed two probable causes: the pilot in command's improper inflight decisions and airframe icing. Improper IFR procedures, low ceilings, fog, and icing conditions were listed as factors.

One wonders if airframe ice was really one of the stronger probable causes. The surface temperature was given as 32° F, and there was a strong, overrunning, southerly flow, indicating a temperature inversion. The aircraft may not actually have entered icing conditions until it was at or below 2,000 feet, and it seems unlikely that it would have collected enough ice in that brief time to affect its flight catastrophically. Moreover, the pilot made no mention of icing problems during the flight, and the aircraft was equipped with deicing equipment. When ice is a factor, as in a temperature inversion, with warm air above, the frost often takes charge only after the pilot descends into icing conditions and shoots several unsuccessful approaches there without climbing back into warm air to shed the ice before another try.

It's hard to argue with a finding of improper inflight decisions when the outcome was as it was in this case, but the decisions are worth examining. The pilot flew toward Albany, even though the ceiling was reported at 100 feet, and he started the approach with the ceiling still reported at that value. The glideslope was out, which set the authorized minimum descent altitude at 400 feet above the ground. Regardless of what the reported ceiling may be, the pilot may fly the approach (down to the MDA), because visibility is the determining factor for minimums. In this instance, with the glideslope out of service, three-quarters of a mile was the required minimum visibility. (The marker was out, then it was reported back on. Either way, the minimum was still three-quarters.) In the last discussion about visibility, the controller gave the prevailing visibility as one-half and the runway visibility as seven-eighths, so it was legal for the pilot to shoot the approach. Many other pilots might have operated the flight in such a manner. Even though the ceiling was reported as 100 feet, it could conceivably have been higher in the approach area, and there is no way to tell without going in for a look. There are times when you shoot an approach that you think may not work out, only to find the airport beckoning when you are flying at the MDA (or decision height on a precision approach). And frequently you find yourself sniffing for asphalt on an approach that looked like a piece of cake from a distance. Weather is really as you find it, not as it is forecast, or, in some cases, as it is reported, because the observation is taken at only one point.

The undeniable fact is that the pilot descended below the minimum descent altitude—he had to in order to hit the ground. That altitude is not supposed to be vacated until the runway or the lights are in sight. Could he have had them in sight from a mile and a half out on final? The runway visibility was given as a mile and a quarter in a special observation taken right after the accident, so there's no way to say that it would have been absolutely impossible for the pilot to have seen the runway before leaving the MDA. However, the 100-foot ceiling would seem to preclude any runway sighting from that far out.

A contributing factor not mentioned in the report could have been a wind-shear-induced increase in the rate of descent. The aircraft was flying with a rapidly decreasing headwind component, which can cause a sinking spell. An application of power should have been adequate to cure this, though.

More IFR accidents happen in the approach phase of flight than anywhere else, and night nonprecision approaches (which lack glidepath information) are the number-one problem. Some pilots must fly such approaches with the illusion that, given a tolerable surface visibility, you can descend gradually until you see something. Forward inflight visibility can be almost nil all the way to the ground, even when visibility measured from the ground is at minimums.

Regardless of what happened, this accident emphasized the hazards of the night nonprecision approach. Ice may have been an added ingredient or may not have had a bearing. The strongest message is: don't ever leave the MDA or DH until the runway (or its lights) are in sight and a normal landing is assured. Be double-conservative at night.—R. L. C.

Investigators often satisfy themselves with the tautological observation that the accident took place because the pilot descended below the MDA. Obviously. But why? Sometimes it's distraction. This pilot was carrying on a casual conversation with the tower when he struck the ground, which suggests that he was not, at any rate, flustered and tense. There is something in the tone of the whole conversation, on the contrary, something about his oblique replies to the controller's concern about the weather, that suggests that the pilot was quite prepared to sneak down, perhaps below the MDA, for a look. If so, it would hardly be unusual: many pilots who publicly preach the inviolability of the MDA admit in private that they often bust minimums. No doubt minimums are busted hundreds of times for every one time that someone gets a scare or gets killed doing it. So the question is not why the pilot descended below minimums: the question is, why so far?—P.G.

17.

"JUST PASSING OVER THE OUTER MARKER NOW . . ."

At 0937 on an early winter morning, the pilot of a single-engine retractable contacted a Minnesota FSS and requested a weather briefing for an IFR flight to Pennsylvania. The reports indicated wide areas of poor weather near Pittsburgh, but forecasts called for some improvement later in the day. The pilot said he'd delay his departure and call back later.

Early that afternoon, the pilot again contacted the FSS for a briefing. He was advised that widespread areas of ceilings lower than 1,000 feet existed at Pittsburgh, with visibility running two miles or less in low stratus, fog, and precipitation. The pilot then requested conditions south of Pittsburgh, including current reports and forecasts for Morgantown and Charleston, West Virginia. This information showed a slight improvement over the general conditions in the Pittsburgh area. The pilot then filed an IFR flight plan to Morgantown with Charleston as the alternate, estimating four hours en route with 5:15 fuel on board. Accompanied by two passengers, he departed Rochester, Minnesota at 1537. He climbed immediately to the requested initial cruising altitude of 11,000 feet and proceeded along the flight-planned course. At 1904, the aircraft was 10 miles southeast of the Tiverton, Ohio VOR, still maintaining 11,000 feet.

The following is a transcript of the recorded radio contacts with the flight, which will be referred to as 46 Echo (HLG = Cleveland Center, Wheeling Sector; MGW = Cleveland Center, Morgantown Sector; 750 = Allegheny Airlines Flight 750; MGT = Morgantown Tower; * denotes contact between ground stations).

TIME: 1905
46E—Cleveland Center, 46 Echo with you at 11,000.
HLG—Four-Six Echo, ident, Pittsburgh altimeter 30.08.
46E—Four-Six Echo.

TIME: 1911
46E—Cleveland Center, 46 Echo, I'd like to leave you to get some weather, please.
HLG—Four-Six Echo, approved.
46E—Roger, and I'll let you know when I'm back.

TIME: 1914
46E—Cleveland Center, 46 Echo, I'm back with you.
HLG—Four-Six Echo, roger.

TIME: 1920
46E—Cleveland Center, 46 Echo, over.
HLG—Six Echo, go ahead.
46E—Ah, 46 Echo, I'd like to change my alternate to, ah, Parkersburg, and, ah, when you run me, ah, I'd like to go Victor 8 after Bellaire and then, ah, to intercept the ILS for Morgantown.

TIME: 1921
HLG—Four-Six Echo, you're cleared via, ah, the Garard Intersection and, ah, direct Bobtown Beacon and, ah, direct and, ah, as far as changing your alternate, sir, why don't you give Morgantown Radio a call and follow with them, sir?
46E—Okay, thanks a lot, then, it will be, ah, Victor 8 to Garard Intersection, ah—to Bobtown Radio Beacon, direct.
HLG—Ah, roger, sir, actually it will be radar vectors for the ILS, but, ah, in lieu of lost radar contact that would be the way.
46E—Okay, that will be radar vectors, ah, to the ILS, thank you.

TIME: 1922
*HLG—Okay, Morgantown?
*MGW—My turn for radar control?
*HLG—Okay, Four-Six Echo. Let me explain to you his routing is slightly changed; he's going down Victor 8 for vectors to the ILS course.
*MGW—All right, you do good work.

TIME: 1923
HLG—Four-Six Echo, Cleveland now on 125.1, good day.
46E—Okay, Cleveland Center 125.1.

TIME: 1924
46E—Cleveland Center, 46 Echo with you level at 11,000.
MGW—Four-Six Echo, Cleveland, squawk code 1100 and ident, Pittsburgh altimeter 30:07.
46E—Four-Six Echo, I'd appreciate knowing how that, ah, inbound flight to Morgantown does on the approach.
MGW—Four-Six Echo, roger. I have the Morgantown weather, if you'd like to copy.
46E—I—I just got it a little bit ago.
MGW—Roger.
46E—Ah, do you have a more recent report than the past 10 minutes?
MGW—Forty-Six Echo, it's ah, just about the last, ah, 10 minutes ago when I got it.
46E—Ah, 46 Echo, roger.

TIME: 1927
MGW—Four-Six Echo, descend and maintain 8,000.
46E—Okay, 46 Echo leaving 11,000 for eight.
MGW—Roger.

TIME: 1928
MGW—Four-Six Echo, say your present heading now, please.
46E—Four-Six Echo, heading, ah, 150.
MGW—Ah, 46 Echo, ah, maintain heading of, ah, zero, correction, make that 100, heading 100, radar vectors to the ILS.
46E—Okay, heading of 100, radar vectors to the ILS.

TIME: 1931
MGW—Four-Six Echo, turn further right now, heading of 130, heading 130.
46E—Four-Six Echo, right to 130.
MGW—Roger.
46E—Four-Six Echo level at eight.
MGW—Forty-Six Echo, roger, continue descent to maintain 5,000 now.
46E—Okay, leaving eight for five, maintain five.
MGW—Forty-Six Echo, roger, next report leaving 6,000, sir.
46E—Okay, report leaving six.
MGW—Roger.

TIME: 1932
46E—Four-Six Echo just coming up on 6,000, descending to five.
MGW—Forty-Six Echo, roger.

TIME: 1933
MGW—Four-Six Echo, continue descent now to maintain 4,000.
46E—Okay, maintain four, I'm leaving.
MGW—Roger.

TIME: 1934
MGW—Four-Six Echo, turn further right now, heading 160.
46E—Four-Six Echo, right to 160.
MGW—Roger, 46 Echo, the, ah, radar indicates you two miles west of centerline the ILS, three and a half miles north of Bobtown Beacon.
46E—Four-Six Echo.

TIME: 1935
MGW—Four-Six Echo, cleared for the ILS.
46E—Okay, 46 Echo, ah, is cleared for the ILS.
MGW—Forty-Six Echo, turn further right now, heading of 180 till intercepting the ILS. Position one mile west of centerline and two miles north of Bobtown Beacon.
46E—Four-Six Echo.

MGW—Four-Six Echo, say altitude leaving now.
46E—Four-Six Echo, leaving four.
MGW—Roger.

TIME: 1936
MGW—Forty-Six Echo, radar indicates you on centerline just south of Bobtown Beacon now; radar service terminated at this time, contact Morgantown Tower 120.0, good day, sir.
46E—One two zero point zero, and I'm north of Bobtown.
MGW—Roger.
MGW—Allegheny 750, ah, the other inbound is one mile south of the Bobtown Beacon at the present time, ah, can you reduce any more, sir?
750—Ah, back a little further now, we show eight miles from Bobtown, over.
MGW—Allegheny 750, that is affirm, the other aircraft is now two miles south of Bobtown.
750—Roger.

TIME: 1937
46E—Okay, Morgantown Tower, 46 Echo, over.
MGT—Four-Six Echo, Morgantown Tower, report the outer marker inbound for Runway 18, wind 180° at three, altimeter 30.08.
46E—Ah, Echo, roger, I'm, ah, just passing the outer marker now.
MGT—Four-Six Echo, roger, cleared to land, current Morgantown weather 200 scattered, measured 500 overcast, visibility two, with light drizzle and fog.
46E—Four-Six Echo.

TIME: 1939
*MGW—Morgantown Tower, how is that 46 Echo doing on his inbound?
*MGT—He's outer marker inbound.
*MGW—Okay, call me as soon as you can; I got that Allegheny 750 just, ah, north of the beacon right now about three miles.

TIME: 1940
MGW—Allegheny 750, ah, Morgantown Tower just advised me the other aircraft just crossed the outer marker.
750—Just crossed the outer marker?
MGW—Allegheny 750, roger, that's what Morgantown Tower said.
750—I just heard somebody inside the outer marker on the other radio.
MGW—Roger.
750—Morgantown Tower, Allegheny 750.
MGT—Allegheny 750, Tower.
750—Ah, do you have that aircraft on approach on the ground yet?
MGT—That's negative.

TIME: 1941
*MGT—Hey, Center?
*MGW—Yeah, go ahead.

*MGT—Can you see that 46 Echo on radar?

*MGW—No, I can't.

*MGT—Four minutes since he passed the outer marker, and I still don't have him.

TIME: 1942

750—Ah, Morgantown Tower, 750, you got that guy yet?

MGT—Seven-Fifty, that's negative.

750—Hoo boy.

There was no further radio contact with 46 Echo. A short time later, however, the airport received a call from a witness who reported hearing an aircraft fly over his home, then an explosion. He was able to fix the time of the explosion with relative certainty at 1938—less than one minute after 46 Echo reported outer-marker passage. The wreckage of the aircraft was located a quarter mile west of the witness' house on a sloping ridge at an elevation of about 1,200 feet; 46 Echo struck some wires and low trees before slamming into the ground. The wreckage remained largely intact and there was no fire, but the three occupants had been killed by the impact.

A detailed investigation showed that the engine was developing normal power at the time of the crash and that there had been no failure or malfunction of the controls or airframe prior to impact. The airplane's landing gear was extended, though the flaps were up. The altimeter was set at 30.07. Avionics aboard included a VHF transceiver, one VOR receiver with ILS localizer function and audio marker beacon, and an ADF, which read 259 kHz; the Bobtown Beacon, located at the outer marker for the ILS 18 approach to Morgantown, has a frequency of 269 kHz. It could not be determined if the marker-beacon receiver was on or operating properly at the time of the crash.

The pilot had performed in competent, routine fashion during the entire enroute portion of the flight—up until the time that he began the approach to Morgantown. A trace of the radar track of the flight after passing Garard Intersection reveals that the pilot responded slowly to course changes, and his voice responses were broken and hesitant. The next indication that all was not well aboard 46 Echo came at 1936, when the radar controller gave the pilot his position as established on the localizer just inside or south of the Bobtown Beacon. The pilot acknowledged a frequency change and then stated his position as still north of the beacon. Neither the pilot nor the controller seemed to notice this discrepancy, and radar service was then terminated.

At 1937, 46 Echo reported to Morgantown Tower that the aircraft was passing the outer marker, although the center controller had just relayed the flight's position to the pilot of Allegheny 750 as being two miles south of that position. Apparently the airplane was still at its initial approach altitude of 2,800 feet msl, only 2.6 miles from the airport and less than one minute from destruction. The MDA for a no-glideslope approach to Morgantown was at that time listed as 1,820 feet, yet 46 Echo struck the ridge at an elevation of 1,200 feet—an altitude actually lower than the elevation of the airport. To have reached this tragic position between the time that 46 Echo reported outer-

116

marker passage and the time that the witness noted the explosion, the airplane would have had to have maintained a rate of descent of 1,600 fpm.

Normal procedure for the pilot approaching for this particular ILS would have been for him to use his ADF to assure the vectored course to the Bobtown Beacon and to have noted the ADF's needle reversal as an indication of passing the outer marker. Since the ADF was not properly set on the beacon frequency, however, it is likely that there was no such indication—though it is possible that a false needle swing well after passing the outer marker led the pilot to believe that he was at the critical position when he was actually more than two miles beyond it.

The official probable cause stated that the pilot had "descended below MDA for undetermined reasons." But what would cause a normally cautious and responsible pilot to tune the ADF improperly, lose his position on an ILS localizer, maintain a rate of descent at least triple the plotted glideslope from the outer marker to the missed-approach point, and descend more than 600 feet below the height of the touchdown zone? (The pilot was instrument-qualified and held a commercial certificate with ASEL and instrument ratings, and his total flight time was 550 hours, 75 of which had been logged during the previous 90 days in 46 Echo.) Realistically, no definitive answer can be given. Perhaps several factors completely unknown to us developed aboard the plane and contributed to the accident, but we must consider one important fact: during this final flight the pilot had maintained an altitude of 11,000 feet for a total of three hours and 14 minutes. No oxygen was available on the aircraft. It is not unlikely that the insidious effects of oxygen starvation clouded the pilot's mind, slowed his reactions, distorted his judgment, and eventually led him into the unseen cloudy ridge.—W. L. T.

In fact, hypoxia, as normally understood, cannot have played a role in this accident. Recovery from hypoxic symptoms is immediate when the oxygen supply is restored. The pilot may or may not have been hypoxic during his long trip at 11,000 feet—it's unlikely that he was seriously so, and there is nothing in the transcript to suggest it—but 10 minutes intervened between his leaving that altitude and his crash. Even if he was short of oxygen during the cruise, he certainly wasn't during the approach.

If hypoxia played no part, perhaps anxiety did. Is there not a hint of worry in the pilot's concern with weather conditions that were not, after all, extremely bad, which contrasts strikingly with the insouciant tone of the light-twin pilot in the previous case, who proceeded casually to Albany with weather well below minimums, with fog and freezing rain to boot? It would be difficult, without having experienced it, to realize the crippling, disorienting effects of anxiety, the nervous haste and aimless repetition of rote movements that it encourages. It is sometimes not all that easy to get spatially oriented on an instrument approach; in this case, having made the error of mistuning his ADF and having allowed the discrepancy between the controller's report of his position and his own radio indications to go unresolved, the pilot had set the stage for a confusing, distracting episode in which he might well make the final, fatal error of misreading his altimeter and think that he was at 2,200 feet when he was in fact at 1,200.—P.G.

18.

DISTRACTING DG

December 9, 1969 was a poor day for flying along the coastal waters of the southeastern United States. On that day, the pilot of a Cessna 210G, N5919F departed alone from his home field at Valdosta, Georgia. He flew IFR to Jacksonville, Florida, where he picked up a business associate and continued IFR to Aiken, South Carolina. Later that same afternoon, the pilot and his passenger proceeded to Columbia, South Carolina, where another passenger boarded the aircraft. The three immediately departed on a return flight to Jacksonville. After a routine ILS approach the aircraft landed, and the first passenger departed.

At 1896, Cessna N5919F again left Jacksonville on an IFR flight plan to Valdosta. The flight plan indicated an estimated time en-route of 1:25, with six hours of fuel on board. At 1900 the Cessna established radio contact with Valdosta Radar Approach Control. A transcription of the relevant recorded communications follows (19F = Cessna N5919F; VAC = Valdosta Approach Control; VLD = Valdosta Flight Service Station; * denotes direct land-line communication between stations).

TIME: 1900
19F—Valdosta Approach, Cessna 19 Fox, 118.5.
VAC—Cessna 19 Fox, Valdosta Approach Control, verify at 4,000, ident.
19F—Four Thousand, ident.
VAC—Cessna 5919 Fox, radar contact six miles northwest Taylor VOR.
19F—Okay.
19F—Ah, Valdosta, 19 Fox. What's your latest weather down there?
VAC—Roger, just received the Valdosta 0000Z weather. Estimated ceiling 400 overcast, visibility two, drizzle and fog. Wind 010 degrees at 11. Altimeter 29.96.
19F—Okay, 29.96, thank you.

TIME: 1901
VAC—Cessna 5919 Fox, for vector to the final approach course, VOR Runway 35 approach at Valdosta, fly heading 270, descend and maintain 2,000.
19F—Okay, 19 Fox, ah, out of four for two on 270.
VAC—Roger.
19F—It sure is smooth up here at 4,000. I'd like to stay a little bit longer if I can.
VAC—Roger, make your descent at pilot's discretion.

19F—Okay, let me know when I'm about 12 miles from Valdosta VOR.
VAC—Roger, wilco.

TIME: 1903
VAC—Cessna 19 Fox, fly heading 275.
19F—Two seven five, and my directional gyro's out. I'm having to go just by the compass.
VAC—Ah, that's, ah, affirmative. I notice you on a southwesterly heading now. Can you proceed direct to the VOR?
19F—Yeah, we're all right. We just made an ILS to Jacksonville at 200 feet.

TIME: 1904
VAC—Roger, Cessna 19 Fox, proceed direct to the Valdosta VOR.

TIME: 1906
19F—Valdosta, 19 Fox. About how far out am I?
VAC—Ah, roger, right now you're 14 miles to the east. However, I don't show you navigating direct to Valdosta VOR. Do you need assistance?
19F—No sir, I show I'm on the 275-degree radial of the Valdosta VOR.
VAC—Ah, roger.
19F—I want you to stay with me, though. My directional gyro is giving me a lot of trouble, and I'm having to reset it about every 10 seconds.

TIME: 1907
VAC—Ah, roger.
19F—And 19 Fox. We'll come down out of four for three now.
VAC—Roger, descend and maintain 2,000.
19F—Roger.
*VLD—Approach Control, I have a 07 Valdosta observation for you.
*VAC—Go ahead.
*VLD—Sky partially obscured, measured ceiling 300 overcast, visibility one, moderate drizzle, wind 010 degrees at 11, the altimeter 29.95.
*VAC—Roger.
VAC—Cessna 5919 Fox, new Valdosta weather. You ready to copy?
19F—Go ahead.

TIME: 1908
VAC—Roger, partial obscuration, measured ceiling 300 overcast, visibility one, drizzle and fog. Wind 010 degrees at 11, altimeter 29.95. This is a special observation taken at 0007Z. What are your intentions?
19F—We'd like to make a pass at it.
VAC—Roger.

TIME: 1909
19F—Valdosta, 19 Fox, about how far am I from the VOR, sir?
VAC—Five miles.

TIME: 1910
19F—Roger.
VAC—Cessna 19 Fox, with your gyro, ah, the way it is, ah, can you accept a turn on to final or will you have to make a full approach?
19F—I'm going over the VOR out of 2,000 and, as soon as I hit it, I'm going to take a heading of 006 degrees.
VAC—Cessna 19 Fox, roger, cleared for a VOR Runway 35 approach to the Valdosta Airport.
19F—One-Niner Fox.

TIME: 1911
VAC—Cessna 19 Fox, ah, did you say you were going to turn directly into the airport after you pass the VOR?
19F—Affirmative.
VAC—Roger, radar service terminated, ah, contact Valdosta Radio on 123.6 for airport advisories.
19F—We'll stay with you for a few minutes.

TIME: 1926
*VAC—Valdosta Radio, Valdosta Approach on the line.
*VLD—Valdosta Radio.
*VAC—Do you have an arrival time on November 5919 Fox?
*VLD—No. I haven't heard from him.
*VAC—He was over the VOR at one-three.
*VLD—Well, he sure hasn't come over. I'll let you know just as soon as I hear from him.
*VAC—Okay.

TIME: 1927
*VAC—Any chance of getting a check on the ramp for that 19 Fox?
*VLD—Ah, he couldn't have got in without us seeing him here. We're awfully low, ah, we'd have heard him too. It's so quiet here. He'll usually call coming in, I can't understand.
*VAC—Are you fairly sure that he's not on the ground at the airport?
*VLD—Yeah.

After no response to repeated calls by both Valdosta Approach Control and the Flight Service Station, Cessna 19F was reported as missing. The wreckage of the aircraft was located early the following morning a quarter mile south of Runway 35. The airplane first hit the extreme tops of some tall trees, breaking off numerous small limbs, and then traveled 140 feet before striking a heavy stand of pine trees, shearing them off some 30 feet about ground level, heavily wooded, swampy terrain. Both occupants were killed.

The official report stated the probable cause of this accident as: "Pilot in command descended below minimum safe altitude in IFR flight conditions." While this may be a substantial evaluation of the obvious, other important factors must be considered as major contributors to the final outcome of the flight.

The pilot was well qualified to conduct IFR flight. He held a commercial certificate and single and multiengine land and instrument ratings; his total flight time was shown as over 5,000 hours, more than 150 of them in the last six months. It cannot be denied, however, that he allowed a deficiency in his aircraft to add to an already difficult problem and that his judgment and procedures while making the approach are open to question.

On the day of the accident, the pilot had performed several hours of actual IFR flight and executed an ILS approach to minimums seemingly without difficulty. The passenger who left the aircraft at Jacksonville wrote the following in his report: "The pilot picked me up in Jacksonville Tuesday morning. We flew to Aiken, South Carolina for a business appointment. Later that afternoon we flew to Columbia and picked up his mother-in-law and flew to Jacksonville. I deplaned in JAX, and he proceeded to Valdosta. I am a private pilot, and everything in the airplane was working normally except the directional gyro; it was spinning. We landed in JAX after six P.M., IFR, with no problems—ceiling was about 300 feet. He checked the JAX and Valdosta weather somewhere around Brunswick, I believe through JAX Center. I do not remember the Valdosta weather other than that it was better than JAX."

As Cessna 19F neared its destination, the pilot was advised of the current Valdosta weather: "ceiling 400 feet, visibility two miles. . . ." The published minimum descent altitude was 400 feet above ground level. From his subsequent communications with radar, it was apparent that the pilot was experiencing some course problems due to the malfunctioning directional gyro. At 1908, the pilot was advised of a special weather observation indicating that Valdosta had deteriorated to well below minimums. In spite of this the pilot elected to "make a pass at it."

As Cessna 19F approached the VOR at 1910, the controller asked if the pilot intended to make a full VOR procedure or would accept a right turn of more than 90 degrees to the final approach course. The pilot stated that he would make the turn directly onto final at 2,000 feet. At this point, the pilot was pressing the approach and allowing the situation to become critical. A more prudent choice would have been to execute the full approach procedure, thus allowing the pilot time to be at the proper altitude of 1,800 feet during the procedure turn and of 1,700 feet over the VOR inbound. This would also have made it possible to be established accurately on the proper radial during those last, most critical moments of the letdown.

By delaying the normal descent from his 4,000 foot cruising altitude he was now less than five miles from the VOR and would intercept the final course at more than a 90 degree angle while still above the procedure-turn altitude. As he crossed the VOR at 2,000 feet, he was still 300 feet above the recommended altitude and only 5.7 miles from the runway. At 1912, the flight was advised, "radar service terminated, contact Valdosta Radio. . . ." The pilot responded by saying that he would remain with Approach Control for a few minutes. This was the last transmission from Cessna 5919F.

What really occurred after passing the VOR inbound nobody will ever know. However, by reconstructing the circumstances surrounding the approach two possibilities can be determined. Crossing the VOR, the pilot would have experienced some difficulty in establishing the final course due to the faulty directional gyro. At the same time he would have been required to maintain a slightly higher than normal rate of descent because of his excessive altitude. His concentration upon these problems might account for the fact that he did not contact Valdosta Radio. As the flight proceeded inbound to the airport, much of the pilot's attention had to be devoted to holding his course. Because of this it is entirely possible that he neglected to monitor his descent to the point that the aircraft passed below the MDA and into the high trees.

Although indications would lead to the conclusion that this accident is only another example of a pilot erring by intentionally ducking under the MDA, other evidence points strongly to an even simpler cause. While the aircraft suffered severe fragmentation, including separation of the wings and empennage, the position of the cockpit controls and instrument readings could be determined. This showed the aircraft to be in a normal landing configuration—gear down and locked, flaps extended 10 degrees, mixture rich, and prop control full forward—except for one thing: the right fuel-tank gauge read "full," the left read "empty," and the fuel-selector valve was positioned on the left tank.

Could it be that the pilot neglected the prelanding checklist during the press of his DG failure and experienced power loss at a critical point in his approach due to failure to check "fuel on fullest tank?" The wing tanks and fuel lines were too badly damaged for this to be determined with any degree of certainty, but the possibility certainly exists.

Whether this or an actual approach-procedure error was the cause, the lesson remains clear. Any unexpected aircraft or equipment deficiency, especially under instrument conditions, must be dealt with particularly carefully. "Stay on the ground" is the only truly safe rule if you find yourself with an airplane even slightly inadequate for the task at hand—a choice that the pilot of 19F had at Jacksonville and at several other points during the day. If the failure occurs in flight, get to VFR conditions and land as expeditiously as circumstances permit. Finally, if the latter choice isn't available, utilize the greatest possible degree of caution in operating under the press of the deficiency. An equipment outage can have a snowballing—and fatal—effect.—W. L. T.

The fuel-starvation theory is, on reflection, not a probable one. The pilot took off from Jacksonville with six hours' fuel in two tanks and crashed two hours later: even in the unlikely event that he never switched tanks during the flight, he still would have had an hour's fuel left in the tank from which he was feeding at the time of the crash.

This one is puzzling: even with a 300 foot ceiling and one mile he should have been able to avoid hitting the ground in level terrain. His position close to and aligned with the runway suggests that, despite the DG trouble, he had executed the approach successfully.

One inevitably comes back to the NTSB conclusion: he simply descended too low. At one's home field, it's tempting to make one's own minimums. If the ceiling was a little lower than reported—ceiling reports are admittedly approximate, especially when they are of the "partially obscured" type—and if he happened to be looking inside rather than outside when the trees loomed out of the fog, no further explanation need be sought.
—P.G.

122

19.

MDA: LAST POINT OF SAFETY

At 1605 on the afternoon of February 6, 1970 a 47-year-old pilot—a technical representative for a large electronics firm—called the Pittsburgh Flight Service Station and requested a briefing for an intended flight to Cincinnati, Ohio. At this time the Great Lakes area was experiencing a normal midwinter weather pattern. A frontal system was moving south from the Canadian border, with a trough of extreme low pressure extending from Illinois across southern Indiana and Ohio and into West Virginia. The entire storm front was characterized by wide bands of low clouds and moderate to heavy precipitation. The pilot was advised of the frontal activity and given the latest sequence report for Greater Cincinnati Airport, which indicated that the ceiling was a measured 300 feet overcast, with visibility one mile in light rain. The pilot then said that he would check the weather later that evening.

At 1707, the pilot entered the FSS office and received a full weather briefing before filing an IFR flight plan from Pittsburgh to Lunken Municipal Airport at Cincinnati. Forty minutes later, the pilot, accompanied by one male and two women passengers, left Allegheny County Airport in Beechcraft Bonanza N726B.

The aircraft had some important limitations, although the pilot was well qualified to conduct this flight—his commercial certificate included ratings in single and multiengine land aircraft as well as an instrument rating, and, while his current logbook showed a total flight time of only 319 hours, of which more than 65 hours was logged as instrument time, he was known to have been a U.S. Navy flight officer for 10 years as well as an active reserve pilot. November 726B was legally equipped for IFR flight, with dual VOR/localizer receivers, dual transceivers, and a three-light marker beacon, but no glideslope receiver was installed, which significantly increased the airplane's maximums for an ILS approach.

The flight proceeded along the planned route without reported incident. At 1911, the pilot was advised by Indianapolis Center to contact Greater Cincinnati Approach Control Radar for the approach-and-landing segment of the flight. The aircraft was nearing Mason Intersection at this time at an altitude of 6,000 feet. The following is a transcript of the relevant radio communications concerning the flight (26B = Beechcraft N726B; CAC = Cincinnati Approach Control; GCT = Greater Cincinnati Tower; * indicates contacts between ground stations on direct land lines).

TIME: 1911

26B—Cincinnati Approach from Bonanza 726 Bravo at 6,000 coming up on Mason, over.

CAC—Bonanza 726 Bravo, Cincinnati Approach Control. Radar contact seven miles northeast of the Mason Intersection, Cincinnati altimeter 30.14, maintain 6,000.

26B—Understand 30.14 altimeter, will maintain 6,000, thank you.

CAC—Bonanza 726 Bravo, Lunken weather indefinite, ceiling 300, sky obscured, visibility one-quarter, light drizzle, fog. And, Bonanza 26 Bravo, that's below minimums for that approach: advise your intentions.

26B—Ah, this is 26 Bravo. What is the weather at Greater Cincinnati?

CAC—Two-Six Bravo, Cincinnati weather indefinite, ceiling 200, sky obscured, visibility three-quarters, light drizzle, fog. Runway 18 visual range more than 6,000 feet.

26B—Ah, this is 26 Bravo. I think I'd rather have a surveillance radar approach to Greater Cincinnati, over.

CAC—Roger, Bonanza 26 Bravo. It would either be a surveillance approach, Runway 18, or you can have an ILS 18 approach, your choice.

26B—I'd like a surveillance approach, over.

CAC—Two-Six Bravo, roger.

TIME: 1918

CAC—Bonanza 26 Bravo, turn right heading 270, vectors for a surveillance approach Runway 18.

26B—Right turn to 270 for surveillance approach to 18, thank you.

TIME: 1922

26B—Cincinnati Approach, Bonanza 26 Bravo. Were you calling, over?

CAC—Two-Six Bravo, negative, say your heading.

26B—Present heading 270, over.

CAC—All right, Bonanza 26 Bravo, you're tracking about 240, turn right heading 290.

26B—Understand right turn to 290, thank you.

CAC—Bonanza 26 Bravo, descend and maintain 4,000.

26B—Descend to and maintain 4,000, leaving six.

CAC—Bonanza 26 Bravo, you're 11 northeast of the airport, descend and maintain 2,700. You're being vectored for a left base, perform landing cockpit check.

26B—Understand 11 northeast, descend to and maintain 2,000, vectors for a left base, thank you.

CAC—That's 2,700, the altitude, 26 Bravo.

26B—Understand 2,700, thank you.

CAC—Bonanza 26 Bravo, turn left heading 270.

26B—Left to 270, thank you.

TIME: 1928

CAC—Bonanza 26 Bravo, if communications are lost for more than 15 seconds while on final, proceed direct to the Addyston Outer Locator and hold, maintain 2,700.

26B—Understand if communications lost, ah, then proceed to Addyston Outer Locator, hold and maintain 2,700, wilco.

CAC—Roger.

CAC—Bonanza 26 Bravo, turn left heading 220.

26B—Left turn to 220, holding at 2,700.

CAC—Roger, prepare to begin descent to your minimum descent altitude in four miles.

26B—Understand prepare for descent in four miles.

TIME: 1929

CAC—Bonanza 26 Bravo, you're eight and one-half final for Runway 18; you need not acknowledge any further transmissions except clearance to land.

CAC—Bonanza 26 Bravo, turn right heading 250.

CAC—Bonanza 26 Bravo, eight miles from end of the runway, heading 250, left of centerline, correcting.

CAC—Bonanza 26 Bravo, turn left heading 230.

TIME: 1930

CAC—Bonanza 26 Bravo, six and one-half from the end, correction, seven and one-half miles from the end of the runway.

CAC—Bonanza 26 Bravo, six miles from the end of the runway, descend to your minimum descent altitude.

TIME: 1931

CAC—Bonanza 26 Bravo, turn left heading 220.

CAC—Bonanza 26 Bravo, turn left heading 210.

CAC—Bonanza 26 Bravo, five miles from end of runway.

CAC—Bonanza 26 Bravo, four and one-half miles from the end of the runway on centerline.

CAC—Bonanza 26 Bravo, four miles from the end of the runway, heading 210.

TIME: 1932

CAC—Bonanza 26 Bravo, four miles from the end of the runway, heading 210, slightly right of centerline, cleared to land Runway 18, wind 220 degrees, 10.

CAC—Bonanza 26 Bravo, turn left heading 205.

TIME: 1933

CAC—Bonanza 26 Bravo, three miles from end of runway, advise approach lights in sight, heading 205.

CAC—Bonanza 26 Bravo, turn right heading 210.

CAC—Bonanza 26 Bravo, turn right heading 215, two and one-half miles from the end of the runway slightly left of centerline, end of the runway, slightly left of centerline.

CAC—Bonanza 26 Bravo, turn right heading 220.

TIME: 1934
CAC—Bonanza 26 Bravo, two miles from the end of the runway, heading 220, correcting back.
CAC—Bonanza 26 Bravo, turn left heading 225.
CAC—Bonanza 26 Bravo, turn left heading 220, one and one-half miles from the end of the runway.
CAC—Bonanza 26 Bravo, one and one-half miles from the end of the runway, heading 220.

TIME: 1935
(Sounds of an open mike hitting metal object)
CAC—Bonanza 26 Bravo, radar contact lost, if the approach lights are not in sight, climb and maintain 2,700, maintain present heading.
*CAC—Do you see the Bonanza? I lost radar with him, no radio contact.
*GCT—Negative.
CAC—Bonanza 26 Bravo, do you read approach control?

No further contact could be established with the flight, and an immediate search was begun. The wreckage was located 7,500 feet from the approach end of Runway 18. Initial investigation showed that impact first occurred when the left wing struck two trees at a point approximately 45 feet above ground level. The right wing then hit several smaller trees as the aircraft continued forward for another 170 feet before hitting the ground. After ground contact the aircraft bounced and slid for another 45 feet before coming to rest inverted.

The aircraft was totally destroyed, but there was no fire. A detailed examination of the wreckage uncovered no evidence of power, structural, or control failure prior to the crash. All systems were in operating condition, and the various controls were found to be in the normal position for approach and landing. The landing gear was determined to be in the down-and-locked position, and the altimeter setting was at the correct pressure reading. The pilot and women occupants were killed in the crash. The male passenger, occupying the right front seat, survived when his seat belt failed and he was thrown clear. Although seriously injured, he could remember most of the flight prior to the crash.

A book of approach plates found in the wreckage was closed, with the outside covers smeared with mud and oil. Upon opening the book to the plates for Greater Cincinnati Airport, it was noted that no mud or oil appeared on these pages, nor could the survivor recall seeing the pilot open the book during the final approach.

At the time the flight left Pittsburgh the Cincinnati forecast indicated that Lunken Airport was expecting ceilings of less than 500 feet, with a visibility of one mile or less. While this forecast was above the required minimums for a full ILS approach, the weather was already at or below the minimums required for an ILS without a glideslope. In spite of this, the pilot did not select an alternate airport for his flight plan.

When the pilot arrived in the Cincinnati area, the Lunken Airport weather had deteriorated below the forecast ceiling and was closed to all IFR traffic. When the radar controller advised him of this fact, a decision was made to attempt a nonprecision radar approach to the nearby Greater Cincinnati Airport, where the ceiling was 200 feet and visibility three-quarters of a mile. This weather was below the requirements for a no-glideslope ILS procedure, the minimum descent altitude being 327 feet agl, but an attempt at this approach would have been legal, due to the visibility reported at the three-quarter-mile minimums and the RVR reported as more than 6,000 feet—well above the 4,000 feet needed for this approach.

When the pilot accepted the radar-surveillance approach to Greater Cincinnati, he may have made an important but incorrect assumption that began the fatal sequence of events that led to the crash. At the time the radar approach was started, the weather minimums were also well below the ceiling requirements for this approach, although the visibility reported still made the attempt legal. In fact, the published MDA is a full 100 feet higher for the radar-surveillance procedure than the MDA listed for a no-glideslope ILS. At the time N726B was cleared for this approach, the reported ceiling was more than 200 feet below the published minimums.

In all probability, the pilot assumed that, since the approach was accepted by the controller, the minimums were above those required for the radar approach. Once under positive radar control, the pilot evidently placed complete dependence on the radar operator to assure the safety of the flight, even to the point that he expected the controller to advise him if he had reached his MDA and missed the approach.

A key point in the approach was reached at 1930, when radar advised: "...six miles from the end of the runway, descend to your minimum descent altitude." The pilot then acknowledged beginning the descent, but at no time during the numerous radio contacts did either the pilot or the controller specify that the MDA was 1,300 feet msl (427 feet agl). From this point on the flight was furnished with headings and the distance to the runway at each half-mile interval, but no recommended altitudes were given throughout the approach.

From the time the final descent was begun until the impact with the trees, less than five minutes had elapsed. During this period the transmissions by the radar operator indicate that the approach was proceeding normally, with only minor course corrections required to keep the aircraft on the final approach course. This segment of an instrument flight is one of the most intense for the pilot: his total concentration is upon the voice of the controller and upon responding to the numerous heading changes if the final moments of the flight are to be conducted safely; however, the pilot must never forget the most important figure of all—the altitude of his minimum descent. The MDA represents his last point of safety and must never be violated.

Numerous recent accidents have been blamed on the pilot's knowingly descending below the minimum descent altitude in an attempt to complete a low-visibility approach. This is regarded as a cardinal sin. Perhaps the only greater sin is to descend below this point without realizing the error. It is doubtful that the pilot of N726B had the opportunity to review his mistake before the sudden impact. The official findings of the probable cause of this accident were listed as: "Pilot in command [performed] improper IFR planning and operations with improper inflight decisions. [He] attempted operations with known deficiencies in equipment, descended below published minimums."

As a final note of irony, we offer this quote from the Terminal Air Traffic Control Handbook, 71.10 8a, dated January 1, 1970: ". . . if recommended altitudes are requested, inform the aircraft that recommended altitudes will be given each mile on final approach down to the last mile which is at or above the published minimum altitude/MDA. Issue advance notice of where descent will begin and, if requested, issue the published minimum/MDA." Since it is obvious and unfortunate that some regulations are more concerned with saving words than lives, it would be wise for any pilot to accept all the assistance provided and request the help that is kept in reserve.— W. L. T.

While it is true that a surveillance approach requires concentration by the pilot, it is actually a less demanding approach than almost any other, partly because the pilot is relieved of any navigational task and partly because he has the emotional comfort of being told what to do by someone else, from whose sangfroid he can draw courage. It is the standard military approach, and it effectively divides the labor of the approach between two people.

Approach plates are published for radar approaches: they contain information about the airport but nothing indispensable—except the minimum descent altitude. The pilot's failure to have the plate binder open is not so puzzling if one assumes that he had previously checked the field elevation or that of nearby Cincinnati Lunken and had planned to add the ceiling to it to get his minimum descent altitude. He had already worked himself into a corner, choosing between airports with below-minimums ceilings and then taking the easy way out by asking for a ground-controlled approach. The MDA gives a good margin for error, but, when one descends below the MDA, the margin evaporates quickly. Approaches to improvised minimums, besides being illegal, require extremely precise flying—more precise than can reliably be managed.

The next account concerns a pilot who, though in a difficult situation, would not descend below the MDA. He consequently had to fly a missed approach with one engine out and a load of ice—an appalling combination of circumstances. He almost made it—and then fate played him a cruel trick.—P. G.

20.

"... CAN'T CONTINUE ON THAT LEFT ENGINE ..."

The airplane, a medium twin, left an airport in a Western state on a late-autumn afternoon on the return leg of a scheduled air-taxi flight. For purposes of the transcript to follow we will refer to the flight as number 374; its destination was Denver, Colorado.

The young pilot held a commercial certificate with single, multiengine, and instrument ratings. He had logged over 2,000 hours, 148 of which were in the aircraft that he was flying at the time of the crash and 252 of which were recorded in the previous 90 days. Flight 374 made one scheduled intermediate stop; the departure was delayed for 30 minutes while the pilot awaited a special advisory on the Denver weather. At 1730 the flight departed on an IFR flight plan to Denver. In addition to the two pilots five passengers were aboard. The aircraft proceeded normally along its assigned routing, and radio contact was established with Denver Approach Control approximately 20 miles northwest of the Denver Vortac at an altitude of 11,000 feet.

The following is a transcript of the recorded radio communications with Flight 374 (374 = Flight 374; DEN = Denver Approach Control Radar).

TIME: 1758
374—Denver Approach, 374.
DEN—Three-Seven-Four, Denver Approach Control, radar contact 20 miles northwest Denver Omni, fly heading 130 at 11,000 vector to 26 final approach course, ceiling indefinite 400, sky obscured, visibility three-fourths of a mile in snow and fog, runway visual range 3,600.
374—Okay, that's 120 and 11,000 feet.
DEN—That's 130.
374—One Three Zero, roger.

TIME: 1801
DEN—Three-Seven-Four, turn right heading 170.
374—Three-Seven-Four. Right 170.
DEN—Three-Seven-Four, turn right heading 190.
374—Three-Seven-Four. Right to 190.

TIME: 1807
DEN—Three-Seven-Four, turn left heading 150.
374—Three-Seven-Four. Left 150, rog.

DEN—Runway visual range now down to 3,000 feet. Does everybody have minimums?
374—Three thousand's good for us.

TIME: 1809
DEN—Three-Seven-Four, descend to maintain 8,000.
374—Ah, 374 is out of 11,000 for eight.
DEN—Three-Seven-Four, turn left heading 050.
374—Left 050, 374, roger.
DEN—Three-Seven-Four, your altitude now and rate of descent?

TIME: 1810
374—Ah, 1,000 feet a minute; we're out of 10.
DEN—Okay.
DEN—Three-Seven-Four, we've got a problem here and a hole has developed —ah, could you increase that rate of descent?
374—Ah, we've got passengers on board, sir, and we don't want, ah, have any pressurization. We can give you an inch more, but that'd be all.
DEN—Okay, don't push it, then, turn right heading 070, and 374, altitude?
374—Ah, we're out of 9,300.
DEN—Okay.

TIME: 1812
DEN—Three-Seven-Four, turn right, make a circle to your right, roll out heading 070.
374—A 360 to the right, 070, roger.
DEN—Three-Seven-Four, altitude?
374—Ah, we're out of 85, and, ah, we got quite a bit of ice on our left engine, might have to shut it down.
DEN—Say again, 374.

TIME: 1813
374—Our left engine's iced up; we can't continue any longer on that left engine.
DEN—Okay, 374, fly heading 120, a vector to 26 final approach course; we'll get you right on in here.
374—Okay, we'd appreciate it, thank you.
DEN—All right.
DEN—Three-Seven-Four, altitude?
374—We're, ah, 8,000.
DEN—Okay, ah, you're 10 miles from the airport, turn right heading 230.
374—Right, 230, roger.
DEN—Three-Seven-Four, do you have the localizer? Turn inbound on it, please.
374—Okay, we will.
DEN—Have you shut the engine down?
374—It's down.

TIME: 1815

DEN—Three-Seven-Four is intercepting the localizer three miles from the marker, turn inbound and cleared for the approach. Ah, the equipment is standing by for you.

374—Okay, thank you, sir.

DEN—Three-Seven-Four, do you have the localizer?

374—We got it.

DEN—Okay, you're about a quarter mile south of it and still southwestbound. Do you want a surveillance approach?

374—Ah, go ahead, give it to us.

DEN—Okay, fly heading 265, 265 is the heading. Maintain 7,000 feet, prepare to begin descent in, ah, 10 seconds; 265 is the heading.

374—Two six five.

TIME: 1816

DEN—Three-Seven-Four, five and a half miles from runway, begin descent to minimum descent altitude, 5,820.

374—Roger, 5,820 feet.

DEN—Three-Seven-Four, turn right heading 270.

DEN—Three-Seven-Four, four miles from runway, turn right heading 275, you're cleared to land, Runway 26 Left, wind 290 degrees, five.

TIME: 1817

DEN—Three-Seven-Four, the runway visual range 3,400, 275 the heading.

DEN—Three-Seven-Four, turn right heading 280, three miles from runway, 280 is the heading.

DEN—Three-Seven-Four, cleared to land, 26 Left. Wind checks 290 degrees at five, runway visual range 3,400, turn left heading 260.

DEN—Two six zero is the heading and one-half mile [sic] from the runway.

TIME: 1818

DEN—Three-Seven-Four, two miles from runway, heading 260, report field in sight, 260 the heading.

DEN—Two six zero, turn right heading 264; 264 is the new heading.

DEN—Three-Seven-Four, 264 is the heading one and a half miles from the runway. Do you have the airport in sight?

374—Negative the airport.

DEN—You have the airport?

374—No airport.

TIME: 1819

DEN—No airport, 264 is the heading, turn right heading 266.

DEN—Two six six is the heading.

DEN—Three-Seven-Four, climb and maintain 7,000.

374—Missed approach, climb and maintain 7,000, roger.

DEN—Three-Seven-Four, altitude?

374—Ah, 5,600.

DEN—Okay, fly your present heading, climb and maintain 7,000 feet.
374—Present heading 240.
DEN—Okay.

TIME: 1820
DEN—Three-Seven-Four, altitude?
374—Fifty-seven.
DEN—Okay, how's your ice?
374—Pretty good.
DEN—Okay.
DEN—Three-Seven-Four, altitude?
374—Three-Seven-Four. Ah, we're at 5,700 and descending 500 feet a minute.
DEN—You're descending?
374—Roger.
DEN—Three-Seven-Four is descending, is that correct?

TIME: 1821
374—All we can hold is about 5,700; we're picking up a lot of ice.
DEN—Three-Seven-Four, turn right, heading 010.
374—Right to 010, roger.
DEN—Three-Seven-Four, maintain as much altitude as you can and you're cleared, ah, are you in a left turn now?
374—Roger.
DEN—Okay, continue the left turn to heading 350.
DEN—We'll start an approach to Runway 35.
374—Ah, looks like we're about to lose it.
DEN—Okay, Lowry's right in front of you. What is your altitude?
374—I'm going to bust it!
DEN—Do you see Lowry? It's right below.

TIME: 1822
DEN—Three-Seven-Four, Denver?

There was no further radio contact with Flight 374, but ground witnesses observed the aircraft approaching the north-south runway of Lowry Air Force Base at an extremely low altitude. As it was about to cross the airport boundary on very short final, the aircraft pitched up abruptly, rolled left to inverted, and crashed.

Rescue units arrived moments later and sprayed the wreckage with foam to minimize the possibility of fire. Two surviving passengers were taken from the aircraft with serious injuries, but the two pilots and remaining three passengers were killed on impact. Detailed investigation revealed that the aircraft first struck a seven-foot wire fence while inverted, then hit the ground 20 feet beyond and skidded for another 138 feet, coming to rest still inverted.

Because of the sudden climb when the plane was seemingly only moments away from a successful emergency landing, unusual care was taken to determine the condition of the control system prior to the crash. No evidence of a control malfunction or failure could be found. An examination of the wreckage also showed that there had been no structural failure prior to impact. A thorough check of the weight and balance of the aircraft was made, including the weighing of each item of baggage aboard. No irregularities were noted, and the takeoff weight was computed to be well below the allowable gross weight. A member of the Lowry AFB security police, the first person to arrive at the accident scene, stated that he had observed a heavy concentration of ice on the wings and fuselage of the plane. The aircraft had been sprayed with fire-retardant foam prior to the arrival of National Transportation Safety Board investigators. No ice was on the wreckage at the time of their arrival.

One of the two surviving passengers was an experienced professional pilot and chief pilot of the flight department of his company. In his written statement he described the sequence of events aboard Flight 374: "The first event of significance occurred during the landing roll-out at Laramie. The runway was wet, with a considerable amount of water on it. During our roll-out we went through a large puddle, creating a spray of water over the left side of the aircraft. Just prior to turning off the runway, the left engine quit. After taxiing clear, the left engine was restarted with some difficulty. The copilot commented, 'The spray of water must have caused it to ice up.'

"Takeoff and climb from Laramie appeared normal. However, at a point I believe to be prior to initial approach at Stapleton, the left engine failed. There was little hesitation, no backfiring, and the manifold pressure dropped at a steady rate without fluctuation. The engine was then feathered, and no attempt was made to restart it. Most of the instrument panel was visible to me, the principal exceptions being the navigational instruments other than the magnetic compass and directional gyro. The copilot periodically checked for icing throughout the flight.

"The air-temperature gauge read a constant 50°, and the crew used very little alternate air. When alternate air was applied, a drop of about one inch was reflected on the manifold-pressure gauge. From my position, one seat behind the captain, I could see no visible ice on the leading edge of the left wing.

"The right engine was operated continuously at 40 inches and 3,500 rpm, except for one time when the captain attempted to increase the manifold pressure to 45 inches, which resulted in the engine missing and backfiring, after which he throttled back. We were unable to climb or even maintain altitude; we could maintain only a slight descent.

"We became contact for the first time just prior to reaching Lowry AFB. We passed Lowry, which was within one-half mile off our left side. The entire landing area was completely visible, with a runway heading within 120 degrees of the heading of the aircraft. The ceiling was ragged in all areas that I could see. Suddenly I heard the captain shout 'wires,' and I was aware he was pulling the nose of the aircraft up. I watched the airspeed fall and instinctively yelled 'Watch your airspeed! Maintain V$_{MC}$!' As the airspeed indicator hit 80 knots, we rolled to the left. That's all I remember."

While it is obvious that the pilot of Flight 374 was facing a difficult situation under the pressure of executing a near-minimum IFR approach with one engine feathered into an airport with an elevation of more than 5,000 feet, the emergency was progressing satisfactorily until the missed approach at Stapleton. At this point the condition of the flight became critical, and the pilot was unable to follow the vector headings given by approach control. Instruction to turn right for an attempted approach to Runway 35 was answered by a continuing left turn, probably due to the failed engine and low airspeed.

In the last desperate moments of the flight an attempt was made to land at Lowry AFB. At this time the pilot reported his altitude as 5,700 feet and a descent of 500 feet per minute. In spite of these extreme difficulties the pilot spotted Lowry AFB and made a wide left turn to begin an approach. But in the final seconds, his struggle was lost. Within feet of the runway threshold, the pilot saw power poles on either side of his course, marked with red warning lights. He reacted by pulling up to avoid collision with the unseen wires, resulting in the stall and crash. The large power poles on either side of the runway carried several lines along the perimeter of the base. Ironically, however, the wires had been placed underground between the span directly ahead of Flight 374 to afford a clear and unobstructed approach to the runway.—W. L. T.

Burying the wire did little good, since wires aren't usually visible anyway, and the customary indication of their presence, power poles, was left in place. The pilot's reaction—to pull up—was the wrong one under the circumstances, but, though it is continually drummed into pilots that in a critical juncture nothing is more important than to maintain airspeed, few would claim that, under the circumstances of Flight 374, they too would not have panicked and pulled up. Even in the last instant there was a chance: cutting power on the good engine would have prevented the loss of control. But it is idle to raise such possibilities: the human mind, in immediate peril of death, cannot be counted upon to work smoothly. Sometimes a pilot accidentally does the right thing, and he is later praised for his coolness and presence of mind. When he does not, heads are shaken and there is discussion of what he should have done. Between victims and survivors there is not necessarily much difference. "Time and chance happeneth to them all."—P. G.

21.

"UH, DO YOU READ?"

The airline captain parked his 747 at San Francisco after an all-night, nonstop flight from Tokyo and was at home for some rest by 8:30 A.M. By 2:30 that afternoon he was up and about, checking on a racing automobile as well as the status of his light airplane. He left for the airport at 3:15, and at approximately 4:15, he was aloft, with three others, en-route to an automobile race. After the race, there was dinner, a ride to the airport, and a two A.M. departure for home.

Once aloft, the pilot called an FSS for a weather briefing. Generally stable weather conditions existed over the area. There were some clouds, but tops were low, reported at 2,500 feet by another aircraft, and, as the airline captain and his companions moved toward their destination, flying VFR, it became apparent that a clearance would be necessary to get down through the clouds. The aircraft had been working Oakland Center for VFR advisories and was handed off to Bay Approach Control. The following transmissions related to the flight of this aircraft, which we will refer to as 29 Victor (OAK = Bay Approach Control; SJC = San Jose Tower; * = direct land-line contact between ground stations).

29V—Bay Approach, 29 Victor is squawking 0500, 7,500, en-route to Reid Hillview Airport, San Jose.
OAK—Two-Nine Victor, Bay Approach Control, roger, ident; Bay altimeter 30.14—uh, don't have the Reid Hillview weather, but San Jose's weather is measured ceiling 900 overcast—visibility 12, winds 350 degrees at five.
29V—Uh, roger, would you say the, uh, overcast?
OAK—Measured ceiling 900 overcast.
29V—Okay, thank you—and the altimeter 30.12 (unintelligible)? Say again the altimeter.
OAK—Three-zero-one-three.
29V—Three-Zero-One-Three, thank you.
OAK—Two-Nine Victor, say altitude.
29V—Two-Nine Victor, uh, still 75, but I'd like to—start down pretty soon, uh—if I don't find some holes up here, I'll need a—localizer approach at San Jose.
OAK—Two-Nine Victor, roger, keep me advised.
OAK—Two-Nine Victor, say altitude.
29V—Uh, 29 Victor, uh, is, uh—just leaving, uh, 75. We're gonna go down to—55.

OAK—Two-Niner Victor, roger—and, uh, Reid Hillview is 11 o'clock, seven miles.

29V—Uh, say again that last transmission?

OAK—Reid Hillview 11 o'clock, seven miles.

29V—Ah, roger.

OAK—Two-Niner Victor, Reid Hillview Airport's at, uh, 10 o'clock, three miles.

29V—Ah, roger, 29 Victor—I'm gonna go a little bit south here. Looks like I might be able to get down VFR. If not, I'll want, uh—localizer approach to San Jose.

OAK—Two-Niner Victor, roger.

29V—Uhh, Bay Approach, this is 29 Victor; uh, I'd like a VOR—or rather a localizer approach to San Jose.

OAK—Two-Niner Victor, roger. Say altitude.

OAK—Two-Niner Victor, say altitude.

29V—Uhh, 29 Victor, uh, five thou.

OAK—Two-Niner Victor, roger, descend and maintain 4,000. Turn right heading 330, intercept the localizer.

29V—Uh, 29 Victor. Would you say again, please?

OAK—Two-Nine Victor, descend and maintain 4,000. Turn right heading 330 and intercept the localizer, or did you want to make it on your own?

29V—Uh, roger, give me radar vectors onto the, uh, localizer.

OAK—Roger, right turn 330 and intercept the, uh, localizer.

29V—Okay, turning to 330.

OAK—Two-Niner Victor, uh, do you intend to, uh, break off the approach and go special VFR to Reids or, uh, land at San Jose?

29V—I'd like to go, uh, Reid Hillview, over—after I get underneath.

OAK—Two-Niner Victor, roger. Position is, uh, three and a half miles from the outer marker, for—29 Victor cleared for localizer approach, Runway 30 Left.

29V—Two-Nine Victor cleared for a localizer approach 29 left.

OAK—Three-zero left.

29V—Ah, 30 Left.

OAK—Two-Niner Victor, roger, contact San Jose Tower—124.0.

29V—San Jose Tower, this is 29 Victor.

SJC—Two-Niner Victor, San Jose Tower, report the outer marker, the airport in sight, and going over the Reid Hillview.

29V—San Jose Tower, this is 29 Victor, do you read?

SJC—Two-Niner Victor, San Jose Tower.

29V—San Jose Tower, this is 29 Victor, do you read?

SJC—Two-Niner Victor, San Jose Tower, how do you read?

29V—San Jose Tower, this is 29 Victor, do you read?

*SJC—If 29 Victor comes back to you, give him, uh, 120.7. He's not hearing, uh, 24.0.

SJC—Two-Niner Victor, San Jose Tower.

29V—This is 29 Victor, go ahead.

SJC—Two-Niner Victor, change to 120.7.

SJC—Two-Niner Victor, San Jose Tower.

*SJC—Where's 29 Victor now?

*OAK—It's a good question.

*SJC—Okay, he wasn't getting me on 24.0, and if he comes back, would you give him 120.7?

*OAK—Okay, I don't see him. I assume he's probably broken off and went on over to, uh, Reid Hillview.

*SJC—Might have—I don't see him anywhere.

*OAK—Ah, I hope so, anyway.

*SJC—Yeah.

*OAK—Was wondering where he was too. He was asking where he was. We don't see him, though.

*SJC—Okay, thank you.

SJC—Two-Niner Victor, San Jose Tower.

*SJC—Bay, San Jose.

*OAK—Yeah.

*SJC—Yeah, I just kept getting all kinds of phone calls. Two-Niner Victor might have crashed. I'm getting, uh, county communications, and what have you, telling me that they have got a crash reported out there.

The reported crash was 29 Victor. The time of the accident was listed as .0322 local time, shortly after the last transmission from 29 Victor was recorded. The wreckage was 1.25 miles from the San Jose outer marker.

The National Transportation Safety Board's findings of probable cause included the following: pilot-in-command—spatial disorientation; pilot-in command exceeded designed stress limits of aircraft; overload failure; separation in flight; and pilot fatigue. The parts about exceeding the designed stress limits, overload failure, and separation in flight were obvious. A portion of the left wing was found 604 feet from the main wreckage, and parts of the empennage, door, and windows were strewn along the path between the left wing and the main wreckage. The airplane came to rest in a residential area, and the consensus of the witness reports was that there was a loud engine roar, swishing noises, the sound of a breakup, and diminishing noise to the time of final impact. The wreckage pattern and the witnesses' statements were all consistent with the high-speed breakup of a clean aircraft after a loss of control.

The question: Why did a 20,000-hour airline pilot lose control of a light aircraft during a rather simple instrument arrival? The NTSB listed pilot fatigue as a miscellaneous condition bearing on the accident. There is no doubt that this could have been a factor. The pilot might have been suffering from "jet lag" after his all-night trip halfway around the globe that had terminated but 20 hours before the accident. He could also have been suffering from lack of sleep, for only six hours had elapsed from the time that he reached his home until he was out inspecting a racing car. However, a person who was with the pilot at the races, who had dinner with the foursome, and who later took them to the airport reported nothing unusual.

At the airport the pilot conducted a rather methodical preflight inspection before starting, and after start he warmed the engine for about five minutes before starting to taxi. This suggests a careful operation by a pilot not in any compulsive rush to get home and get to bed. Some of the communications do suggest the possibility of fatigue, though, for the pilot did have to request repeats of simple instructions from the controller. To the fatigue possibility add a troubling diversion—the communication problem with San Jose Tower while maneuvering for the approach. Then ponder the fact that the pilot was current in a 747 but did not have extensive recent IFR night experience in light aircraft, and it is possible to collect enough straws to break 29 Victor—even with a 20,000-hour pilot at the controls.

When flying single-pilot IFR, it is necessary to assign the proper priorities to things, and, in all events, control of the airplane comes ahead of communications. In this case, no great importance could be assigned to anything that the man on the ground could or would have said at the time that 29 Victor was trying to make contact. The aircraft had been cleared for the approach, and the pilot could have continued until he broke out beneath the clouds without calling the tower and without bothering the controller. In fact, when 29 Victor was a no-show at the airport, two controllers had some routine conversation about the fact that he might have broken out and headed for Reid Hillview. The pilot, though, was devoting time to communications—he called five times, with the last call made only seconds before the aircraft broke up in flight. The communications problem presented itself at the worst possible time—just about the point that the pilot descended into the clouds in the approach.

Lessons: putting aircraft control before all other chores is number one on the list. Next comes recognition that current experience in large aircraft is no more valuable in flying small ones than is small-airplane experience in flying a 747. Lastly, operating behind the freshness curve can put even the most skillful and experienced captain at a disadvantage.—R. L. C.

After a series of weather-related accidents in which the choices and actions of the pilot played an important role, the next case is one in which the catastrophe came with the brutal and arbitrary suddenness of a hammer falling.—P. G.

22.

THE THUNDERSTORM SPIRAL

At 1453 Greenwich mean time on a July morning, an instrument-rated pilot called the National Weather Service office and requested a weather briefing for a proposed flight from Huntsville, Alabama to Daytona Beach, Florida. The Weather Service specialist on duty gave the pilot the latest weather, which was 700 broken and five miles' visibility, and told him that it was IFR but that the forecast called for improvement to VFR within two hours. The latest radar summary chart indicated thundershower activity along the west coast of Florida but none along the route to Daytona Beach. The NWS briefer added that scattered thundershowers were expected to develop in southern Georgia and Florida during the afternoon. Apparently he made no mention of the surface weather chart, which showed a quasistationary front extending southwestward from eastern Kentucky to northeastern Texas and a squall line extending south-southwestward from northern Mississippi to the Gulf of Mexico, well west of the pilot's route of flight. Nothing on the surface chart suggested any problem, but the 500-millibar (upper-air) chart showed a low-pressure area with colder temperatures aloft, indicating that there would be some instability along the route of flight. The upper-air chart was not mentioned to the pilot, nor was it mentioned in the statement given to the National Transportation Safety Board by the NWS regarding the weather of this particular morning.

The pilot filed an IFR flight plan with the FSS at 1545Z. No weather information was requested or given at that time. At 1623Z, the pilot called ground control for taxi instructions and a clearance. The transaction was routine, although the controller did have to read a "cleared-as-filed" clearance to the pilot twice. The aircraft was airborne at 1626Z. The climb to altitude was restricted, with clearance to the planned altitude granted only after the airplane was in contact with Atlanta Center. At 1651Z, the pilot reported to Atlanta Center that he was level at 7,000 feet. The flight proceeded normally for the next few minutes, with nothing more than the usual frequency changes, "level at seven," and requests to squawk. At 1724Z, the pilot was asked to take up a 180-degree heading for vectors around other IFR traffic. The following transcript covers the next 24 minutes of the flight of the aircraft, which we will refer to as 23 November (ATL = Atlanta Center; JAX = Jacksonville Center; * indicates direct land-line contact between ground stations).

23N—Two-Three November. I was on my last frequency with Atlanta Center; I was given, ah, around traffic, and I didn't know, ah, to resume normal navigation or not.

ATL—Two-Three November, continue 180 heading, vectors to Eufaula.

23N—Two-Three November. Thank you.

ATL—Two-Three November, resume navigation direct Eufaula, Victor 159.

23N—Two-Three November.

*ATL—Albany Low, two-five handoff.

*JAX—Albany Low.

*ATL—Two-Three November is three miles southeast of Eufaula 7,000.

*JAX—Seven southeast.

*ATL—Seven southeast.

*JAX—Ha, let's check another one here.

*ATL—Do you see him?

*JAX—Just terminate him; I don't see him.

ATL—Two-Three November, radar service terminated, contact Jacksonville Center 128.1.

23N—Two-Three November. One two eight point one.

ATL—Roger.

Everything was completely routine through the time of the last transmission at 1748Z, yet, unknown to the controllers, the airplane that appeared to be flying normally at 7,000 feet was about to be destroyed in a thunderstorm. According to the National Transportation Safety Board report, the accident took place at 1748Z—the time of the last transmission.

Much conversation ensued between controllers when the aircraft did not report on the new frequency, one bit of which is very pertinent: "I talked to Navy Albany and they looked out there on radar between Albany and Sawmill. They said they can't see anything, even a primary, and we have a lot of weather in that area . . ." That mention of weather was the first in the entire transcript. The pilot had made no comment on conditions, nor did he request weather information, and no controller mentioned it. The lack of talk about weather didn't signify a lack of it, though, because weather-radar reports for the area showed a broken area of thunderstorms, tops to 45,000 feet, beginning at about the point where 23N met instant disaster.

The briefing, obtained almost three hours before the accident, indicated no problem along the route of the flight—only scattered thundershowers forecast for an area farther down the line. Nobody could have looked over the pilot's shoulder as he planned the flight and suggested that the level of risk would be high. Even if the pilot had requested updated weather information when he filed a flight plan an hour after the initial briefing, the information available at that time probably would not have prompted a cancellation of the flight.

As the airplane moved along at 7,000 feet in the initial phase of the trip, the weather must have appeared cooperative. About the time that the plane passed Columbus, Georgia, 40 miles north of the accident site, it was reported 3,000 scattered and 25,000 scattered. The pilot made no request for altitude changes, nor did he mention any turbulence, so it is reasonable to assume that everything was serene aboard 23N up to the last minute.

The radar report clearly indicates that the thunderstorms were just northeast of the low-pressure area showing on the 500-millibar chart and that the accident occurred just as the airplane entered the area of thunderstorm activity. A pilot report from a U.S. Air Force C-131 flying northwest of the accident area gives a further picture of how it looked: "Over Tuskegee at 1758Z in the clear, negative turbulence, negative ice at 9,000. During climb from Maxwell AFB in and out of clouds, bad to the southeast of Maxwell; need radar to go 20 miles northeast of Tuskegee, line 200 miles long, tops 15,000 to 20,000, building."

Neither the situation nor the observations nor the pilot report suggests that the storm would have been embedded in other clouds, so there is a strong probability that some evidence of the thunderstorm would have been visible to the pilot well in advance. Why would a pilot fly straight into a thunderstorm? Perhaps he felt that the controllers would advise him if any weather appeared across his route of flight. That was an error on the pilot's part. Some controllers are good about telling pilots of weather; others have to be asked. One should always ask, especially if the sky becomes at all menacing. This pilot might also have been complacent because of a weather briefing that indicated good weather, but a weather briefing in no way compares with what is visible outside the windshield. These days many pilots will let the responsibility for weather information lie with the Government employee on the other end of the phone (or radio). But in doing so they may learn the true meaning of "pilot in command" the hard way. When the weather scene—the picture in the windshield—started changing from that which was forecast, the controller would doubtless have responded to a question about activity along the airway if only the pilot had asked. The pilot could also have switched over to a flight service station and asked. If the pilot had listened to the ADF, the static would have indicated thunderstorm activity.

Once he misjudged the weather, it was all over. Even that needn't have been so, however. A thunderstorm is a terrible place to be in any light airplane, but it isn't necessary to consider the first sharp jab of an updraft to be fatal.

There are two basic ways in which airplanes come to grief in thunderstorms. They can suffer structural failure, caused by vertical gust-induced overstresses on entering the turbulent area, or the pilot can lose control of the aircraft in the turbulence. When he loses control, the plane can enter a spiral dive or simply fly fast and furiously into the ground. Improper dive recovery often causes the structural failure. This airplane apparently did not fail as it first entered the storm area: investigation showed that the parts of the tail and wing separated, but they did so at a relatively low altitude. It appeared to be a typical loss-of-control ending.

Several factors play on a pilot flying instruments in extreme turbulence. Fear is a big one, but most pilots fear for the strength of the airframe when they should really be more concerned with their ability to keep the airplane flying at an attitude and speed that will maximize chances for success. Many more airplanes break after a loss of control than on first encountering turbulence. If the wings remain reasonably level and the pitch attitude is held steady enough to yield an average airspeed near the plane's maneuvering speed, the plane will probably make it through the storm.

Instead of maintaining a level attitude and reasonable speed, the pilot of the ill-fated flight must have allowed the airplane to wind off into a spiral dive. He may have even attempted a retreat back out of the area of turbulence, which could have hastened the entry into the spiral. If so, less than a quarter of the turn was completed, for the airplane had been following an airway with a 115-degree bearing and broke up just south of it on a southerly heading.

An airplane in a spiral dive is in a precarious situation, but it's not an impossible situation if the pilot understands what is happening. The angle of bank will average about 60 degrees, the nose-down pitch attitude about 25 degrees. Once in the spiral, the airplane reaches a stable situation, an equilibrium. The natural forces trying to steepen the spiral are equalized by the forces trying to make the airplane fly out of it. The airspeed will be high, much greater than the speed at which the airplane had been trimmed to fly in level flight, but airplanes are strong enough to survive such maneuvers. The overstress possibility in a spiral occurs when the pilot begins the recovery, whether on instruments and following the proper unusual-attitude procedures or out of the clouds and attempting a VFR recovery. We are taught to recover by leveling the wings first and then correcting the pitch attitude to level. (Trying to bring the nose up first before leveling the wings would only tighten the spiral.) The pilot's instinct naturally leads him to roll out of the bank and pull back on the wheel to bring the airplane to level; instinct serves a pilot poorly in this situation, though, because, for the preservation of life, liberty, and the airframe, the pilot should actually apply some forward pressure on the wheel as the wings are leveled in the recovery from a diving spiral. Why? Because the plane in the spiral is going much faster than it's level-flight trim speed, and, when the angle of bank is reduced, the airplane is going to try to zoom upward of its own accord, putting great stresses on the frame. If the pilot joins in by pulling back on the wheel even more, the overstress possibilities are almost absolute, and the airframe will fail.

Every airplane has a part that will be the first to let go in a given set of circumstances. On many, the horizontal tail fails downward from negative loading, because there's an extreme download imposed on the horizontal tail by the upward deflected elevator. If a pilot is able to manage the stresses with some forward pressure on the wheel and has room to make a gradual recovery from the dive before reaching the ground, there might still be some chance for success. Success eluded the pilot in question, though, and the accident that resulted was a carbon copy of many other accidents about which there is no mystery. A meteorologist can explain the fury of a thunderstorm; an aerodynamicist can tell about the airplane's behavior; and the two together explain why that airplane did not reach its destination. All of which serves to illustrate why there needs to be a touch of meteorologist and aerodynamicist in every pilot.—R. L. C.

23.

TERMINAL COLLISION AREA

The light twin was on an instrument training flight, with an instructor in the right seat and the pilot in the left seat wearing a hood. The skies were clear, and the visibility unlimited. Multiple approaches had been requested at Denver's Stapleton Field; two had been completed, and a third was in progress as the following transcript begins (774 = the twin; DEN = Denver Approach Control; 103 = Rocky Mountain Flight 103; portions in parentheses were not clear on the recording but represent the best interpretation possible under the circumstances).

TIME: 0932
774—Seven-Seven-Four, level eight.
DEN—Seven-Seven-Four, roger, start the base in about four miles.
DEN—Seven-Seven-Four, turn right heading 010.
774—Zero-one-zero.
774—Seven-Seven-Four's on course.
DEN—Seven-Seven-Four, I'm sorry about that; cleared ILS back course Runway 8, Right approach. You're seven miles from Broadway. Tower 118.3 at Broadway.
774—Seven-Seven-Four.

TIME: 0936:55
774—(unintelligible) Mayday, Mayday, 774 just hit a bird.
DEN—Seven-Seven-Four, uh, what's wrong?
774—Just hit a goose.
774—The left wing is in bad shape; we just hit a bird; we're heading straight in; we got full power (unintelligible).
DEN—Okay, 774, maintain full power and, uh, continue straight in. Stand by this frequency.
774—We're never going to make the runway.
774—You got a helicopter, possibly? We got, a, bad damage here.
DEN—You say you can't make the runway?
774—We got a bad crushed-in left wingtip. You got a helicopter? We're going to try to try to make it in on a road here.

TIME: 0937:35
DEN—Okay, try to make it on the road, okay. We'll get a helicopter in that position right now.

774—We're going to ditch in a lake. We're going to ditch in a lake. (Is everybody all right?)
DEN—Okay, understand, ditch in the lake. We'll get a helicopter right out there.
774—We're squawking 77.
DEN—We'll get a helicopter right out, 774. Just do the best you can. Wind now at Stapleton 150 degrees at seven.

TIME: 0938:35
774—Okay, heading toward the lake on final here.
DEN—Seven-Seven-Four, roger. We're in contact with a helicopter. There'll be someone out.
774—Okay, we think we got it made.
DEN—Okay, real fine, 774, good luck to you.
774—(Thanks a lot for everything.)
DEN—Okay.
774—(unintelligible).
DEN—I can't read you, 774.
774—(Right way.)

TIME: 0942:50
DEN—Seven-Seven-Four, 774, this is Denver Radar, do you read?
DEN—Rocky Mountain 103, you might put an eye out over Sloan's Lake; we just had an airplane going down over there, and you can tell me if he's still afloat.
DEN—Rocky Mountain 103, didja see anything there at Sloan's Lake?
103—Yeah, his vertical stabilizer's sticking out of the water. Part of the tail, fuselage, both wingtips are out of the water, and there are two or three boats around it.

TIME: 0945
DEN—Okay, real fine, sir.

During an early phase of the accident investigation, the instructor pilot stated that he was watching the altimeter subsequent to the receipt of the approach clearance to determine if the pilot under the hood would descend to 7,100 feet over Broadway as stipulated on the approach chart. He noted the altimeter at 7,900 feet an instant before the trouble. The student pilot, who was wearing a hood, stated that he saw only a blur that appeared to be a large goose. He removed the hood, tossed it and the clipboard into the rear seat, and then transmitted the Mayday advising that they had hit a goose. He observed damage to the left wing and cowling. Shortly after the Mayday call the instructor pilot advised the student that he thought that they had struck another aircraft. Both pilots had to apply full right aileron and right rudder to control the aircraft. The lake landing was made with gear and flaps up, and both throttles were closed and the prop controls put in the feather position prior to touchdown. Both pilots evacuated after the water landing and were picked up.

The collision was indeed with another aircraft, not a goose, and the occupants of that aircraft were not as fortunate as those of the twin. One witness stated that the other aircraft, a single-engine retractable, had made a sharp turn to the right a split second before the collision, and the reconstructed wreckage of the two aircrafts tends to verify this. The twin's left engine contacted the bottom of the left wing of the retractable at about the point where the aileron and flap meet, and one tail surface of the retractable made contact with the left wing of the twin as the airplanes merged and then separated. The outboard portion of the retractable's left wing separated, and the airplane crashed out of control. The controller did not see the retractable on his scope and thus did not issue the aircraft as traffic to the twin. The collision occurred one mile outside the inner ring of the Denver Terminal Control Area; there was another midair in the Denver Terminal area a week later. The National Transportation Safety Board's finding of probable cause referred only to the pilots of the aircraft for failing to see and to avoid each other. The accident provides a springboard for discussion beyond that narrow finding, though, and several lessons can be drawn from it.

The first has to do with the twin's mission. There have been a number of midair collisions involving aircraft on instrument training flights, and logic does suggest that there is a high collision risk involved in this type of flying. Instrument training is generally done in a terminal area, often a busy one. One pilot in the aircraft is hooded and is branded as a cheater if he or she peeks outside. The other pilot, the instructor, must teach, and to teach you have to know what the student is doing. This means watching the instruments much of the time. Furthermore, the view to the left often can be partially obstructed by the student's hood. In total, the instructor is diverted and at a slight visual disadvantage when looking for other traffic. The point is to recognize this and to develop a methodical system of devoting time to an active scan for traffic. Another remedy for the instrument-training collision problem might be a third person in the airplane as a traffic observer. Some schools have a second student along in the rear seat as an observer of the flight, and this might help some, but it would be better for the rider to spend his full time searching the sky for other aircraft rather than trying to keep up with what is going on in the cockpit.

Another lesson has to do with terminal control areas. The retractable in this case was a mile outside the inner ring of the TCA and 100 feet below the base of the next ring. Legally a miss is as good as a mile when attempting to avoid a TCA, but one should be especially alert when flirting with its edges, because where there is a TCA, there is a high concentration of traffic. The TCA also tends to compress VFR traffic into a small area beneath the outer rings, and a pilot coming in from a long cross-country, as the retractable pilot was doing, might find it best to fly a few extra miles and give the TCA more berth. Certainly, being near a TCA should be fair warning to keep the head up and on a swivel.

This accident also tells a tale about radar traffic advisories. These are often of questionable value, because the controllers are required to give traffic advisories (to VFR or IFR aircraft) only on a work-load-permitting basis. Moreover, they often do not see aircraft without transponders. As a result, pilots who tend to breathe a little easier and relax their vigilance during radar contact are making a very serious mistake, whether en-route or in a terminal area. When there's inflight visibility, the collision-avoidance responsibility in relation to VFR aircraft belongs solely to the pilot, and being in radar contact should be taken as a fringe benefit of questionable value. Look for the traffic that is called, but by no means assume that even a high percentage of the traffic will be called.

The final visual lesson concerns motivation. Looking for other airplanes is dull, no doubt about it, because you seldom see another one close enough to matter. It's like playing golf for a year before hitting the ball. This is an important flight skill, though, and should be developed to the fullest extent by all pilots. Heads up; scan the sky up, down, and all the way round on a systematic basis, pausing every 20 or 30 degrees to let the eyes take it in and to let peripheral vision work. If the airplane has blind spots, work around them. S-turn, if you must, to make certain that the airplane is not plunging through air that you haven't scanned. Most collisions occur near airports, with final approach the most lethal area. The closer you get, the more effort you should put into looking for other traffic.

One final item to contemplate: the pilot of the twin switched his transponder to 7700, the emergency code, after the collision. This seems to be the appropriate thing to do, for swapping pieces with another airplane in flight is about as emergency as you can get. But, if you are already in radar and communications contact and unless otherwise instructed, it is probably best to remain on the assigned transponder code that the controller is already tracking. In one emergency situation elsewhere, radar contact was lost when a pilot switched from an assigned code to 7700, with the result that the airplane was not tracked for its last minutes of flight.—R. L. C.

People tend to moralize that "if they had been looking out, this wouldn't have happened." The assumption is that, if a collision took place, they must not have been looking out. The grim fact is that you can look and look and still fail to see: under many types of circumstances scanning the sky gives no assurance of seeing an approaching airplane. Furthermore, some of the time, one has to be attending to matters inside the cockpit: it simply isn't possible to be looking in all directions all the time. Everyone depends to some extent upon the unlikelihood of a meeting of two tiny objects moving randomly through a huge space; most of the time, probability is as good a collision-avoidance system as any. Experience often makes pilots casual. After years of not always looking for traffic and never getting hit, one may begin to feel that there is not much point in looking out at all. The last two cases in this collection involve airline pilots on airline flights. Both involve collision with terrain while flying at low level in poor visibility or no visibility. What is striking to begin with is the perfect coolness and unconcern of the crews.—P. G.

24.

"AT LEAST WE SAW THE RUNWAY THAT TIME"

At approximately 1357 Bering Daylight Time on August 30, 1976 Wien Air Alaska Flight 99, a Fairchild F-27B, crashed into Sevuokuk Mountain, Gambell, Alaska while attempting to land at Gambell Airport. The airport is located near the northwestern tip of a three-mile-wide point extending out into the Bering Sea. Sevuokuk Mountain is on the northeast side of the point. The weather at the time of the accident was characterized by low ceilings and poor visibility. Sea fog was reported to be moving in an easterly direction across the airport and toward the mountain. The following transcript is taken from Flight 99's cockpit voice recorder. It begins about 25 minutes before the accident. The times given are approximate Bering Daylight Time (CC1 = a voice identified as that of the captain; CC2 = a voice identified as that of the first officer; and CC3 = an unidentified voice. In all instances, CC denotes cockpit conversation. R1 = radio transmissions by the captain; R2 = radio transmissions by the first officer; CGS = the airline's company ground station at Gambell; ellipses denote words that are either unintelligible or not pertinent; questionable text is enclosed in parentheses; National Transportation Safety Board editorial insertions are in brackets).

TIME: 1331
R2—Roger, should be there in about another 12 minutes. Uh, could you give me your weather, please?
CGS—Roger, roger, stand by.
CGS—Ninety-Nine, Wien Gambell.
R2—Ninety-Nine, go ahead.
CGS—Roger, our current weather is partial obscured sky, visibility one to one and a half miles (just a minute) . . . one mile to half mile with . . . and fog; temperature is 47, wind is calm, altimeter 29.70.
CC2—Wind calm?
CC1—Mile and a half mile, she said [sound of laughter]?
CC2—I thought she said a mile to a mile and a half.
CC1—Huh?
CC2—The first time she said a mile to a mile and a half.
CC1—Oh, did she?
CC2—And then she gave me a mile to a half mile.
CC1—I thought she said a half mile—it's hard to tell.
CC1—Anyway, we're going, anyway.
CC1—Call them.

R2—Understand, uh, the visibility a mile to a mile and a half?
CGS—Roger, roger. Sometimes about five miles; it's very hard to tell . . .
CC1—[Conversation about three-wheel Hondas to be picked up].

TIME: 1334
CC1—Looks like we got fog hanging in there. That's what the problem is.
CC1—We'll take a shot from this way, and, if we can't make it, we'll come around and hit it from that side. If we can't make it from there, we'll turn out and (come in underneath it).
CC2—If we can't make it that way, we'll go into a hover . . .
CC1—We got low fog in here—that's what it is . . .
CC2—Yeah.
CC2—It was up on top of the hill this morning—probably just moved down the hill.
CC1—. . . up on the hill, down below it.
CC2—I (don't) think we can make it thisaway.
CC1—Read the checklist.
   [Approach checklist read]
CC1—I'll tell you what I'll do—just head out this way and go around and intercept the, uh, the 335 radial back in.
CC2—We could land up on the hill there and toss the stuff down the hill to them.
CC1—Could do that.
CC1—That's where they need the airport—up on top of the hill.
CC2—Yeah.

TIME: 1338
CC1—That . . near on the ground.
CC2—What?
CC1—Stuff's darn near right on the ground.
CC3—(going down.)
   [Sound of power reduction]
CC3—. . .
CC2—Which way you comin' in from?
CC1—Huh?
CC2—What way you going to come in from?
CC1—Well, we'll take a look at it this way, then, ah, then take a look at it coming from the other way.
CC3—(Not even) . . .
CC1—Trouble is, the beacon's downtown.
CC1—Have to stay off to the right.
CC2—One-fifty-five inbound?
CC3—(I'm ready for the checklist right now.)
   [Sound of power reduction]
CC1—Put the gear down anyway.
CC2—. . .
CC1—Sixteen flaps and final check.
   [Sound of gear extending]

CC1—(Interception).

CC2—. . .

CC2—Over the water.

CC1—Over the shoreline now.

CC2—We are?

CC1—Yeah.

[Sound of additional power reduction]

CC3—. . .

CC1—No, I can't see it from here.

CC2—. . .

CC2—The beacon's on the end of the runway.

CC1—. . . at the end of the runway.

CC1—Can't get much more lined up.

CC1—Okay, we passed over it. Yeah.

CC1—Raise the gear.

[Sound of gear retraction]

CC3—. . . flaps . . .

CC1—If we can't do it IFR, we'll do it VFR.

CGS—Flight 99, Wien Gambell.

R2—Roger, 99. We're going to try it again there.

CGS—Roger, roger, Estimated time was supposed to be 05 or 15?

R2—What did you say?

CGS—Roger, we're having lots of trouble with our transmitter, and I didn't think I had your on estimated time to Gambell, right. I have it 15; was it supposed to be 45?

R2—No, it was supposed to be 15 minutes, but we're going to be awhile. We're coming around now to try it again.

CC2—Roger, roger . . .

TIME: 1345

[Gambell ADF identifier heard in background]

[Unidentified noise or possible voice]

CC3—. . .

CC1—There's the point over there.

CC3—I think we'll be all right.

CC2—Final checklist still complete except for the gear and flaps.

CC1—Okay.

CC3—. . . take a look.

CC2—. . . want to try it again?

CC1—Huh?

CC2—(Going to try it again from around this way, or you going to try it again from this way here?)

CC1—If I can get that beach line, we'll have it made.

[Sound of windshield wipers activating]

CC1—Watch for the beach.

CC2—Yeah.

CC2—As soon as I see it, I'll let you know.

CC1—(Okay, I've got it here.)
CC1—There's the . . . runway right there.
CC3—Try it again.
CC2—Okay, I've got the village from here.
CC1—Yeah, I've got the runway.
CC1—Sixteen flaps.
CC1—Gear down.

TIME: 1352
CGS—Ninety-Nine, Wien Gambell.
R2—Stand by, we're coming back around again.
CGS—Roger, roger.
CC1—Twenty-Seven.
   [Sound of gear warning horn]
CC2—. . . gear's down, all three, pressure's up, final check complete, final flaps
to go.
CC1—Twenty-Seven.
CC2—You're at 27.
CC2—One-ten.
   [Sound of power reduction]
CC2—Hundred feet.
CC1—Full flaps.
CC2—Got the runway in sight?
CC1—Yeah.
   [Sound of laughter]
CC1—. . . I think we're too late.
CC2—. . . flaps up?
CC2—Yeah, flaps 16.
CC2—Flaps coming to 16.
CC2—Positive rate.
   [Sound of power increase]
CC1—Gear up.
   [Sound of gear retraction]
CC1—We'd never stop.
CC2—What?
CC1—We might have stopped.
CC3—(Off) the end of the runway.
CC2—Think we can get it from that side.
CC1—Huh?
CC2—I think we can get it from that side, that 155 inbound.
CC2—We're coming up on 400 feet.
CC1—Flaps up.
CC1—. . . hard to see, I'll tell you.

TIME: 1355
CC2—I think that 155 inbound is the better way to go.
CC2—At least we saw the runway that time.
CC1—Huh?

150

CC2—At least we saw the runway that time.
CC2—(I can turn them on in just a second here.)
CC1—Just keep them in auto—that's all right.
[Sound of power reduction]
CC2—One-fifty-five heading inbound . . . perfect . . .
CC2—Okay we're over the beacon.
CC1—Beacon right now, final flaps, gear down.
[Sound of gear warning horn, sound of gear extending]
CC2—You're a little fast.
CC1—Yeah, I know.
CC2—. . . (going down)
[Clink]
[High-pitched whistle for one second, exclamation, sound of impact]

During the NTSB investigation, some survivors indicated that they thought that the airplane actually touched down on the next-to-last approach. On the last approach, witnesses reported that the visibility ranged from a quarter of a mile to a mile and a half, and some witnesses reported hearing an increase in engine sound shortly after the aircraft disappeared into fog and just before the impact. Some survivors agreed with this observation, and some of the passengers who could see the ground and were aware of their position relative to Gambell realized that the aircraft was going to hit the mountain and braced themselves.

The NTSB determined that the probable cause of the accident was the flight crew's failure to adhere to prescribed company instrument-approach procedures while attempting to land in adverse weather. The NTSB noted that it was evident that the captain continued to fly in and out of instrument flight conditions in an area and at altitudes where visual flight was mandatory. The NTSB also considered the possibility that the pilots, in their efforts to sight the runway, were not devoting full attention to monitoring directional-instrument indications. Instead, the NTSB believes that they were trying to maintain visual flight during the landing attempts. According to the NTSB, the information from witnesses, survivors, and recorders indicated that the captain did not follow the prescribed instrument-approach procedures during any of his several attempts to land.

Toward the last, the first officer stated that a 155-degree heading inbound would have been perfect, but—there was no evidence to indicate that the captain turned to that heading. He was probably relying on visual clues at that time, and the final turn was stopped on a heading of about 140 degrees, where it remained until impact.

As an aside, it is noteworthy that a majority of the occupants of the aircraft survived this accident. Despite the fact that the high-wing F-27 doesn't particularly look like an airplane that would survive a sudden stop, this design does indeed have a remarkable record of doing so. A sports team, many of whom survived the crash in the Andes described in the bestseller *Alive*, had gone down in an F-27; in other accidents these aircraft have been left on mountain ledges and in forest hollows without total loss of life.

The accident at Gambell involved an air carrier, but there are important lessons here for all pilots, because a large percentage of general-aviation weather-related VFR and IFR accidents are of this very type—flight into terrain with the aircraft under control. In IFR accidents the pilot generally does the deed while maneuvering to land, as in this case, and trouble usually occurs when the pilot deviates from the published instrument-approach procedures.

Approach procedures are formulated to provide minimum acceptable clearance over terrain. The word "minimum" must be stressed, because the criteria for approaches don't allow for error below those altitudes. When flying the last altitude that is part of an approach—the minimum descent altitude in the case of a nonprecision (nonglideslope) approach—the rule is to not leave that altitude unless the approach threshold of the runway, the approach lights, or other markings identifiable with the approach end of the runway are clearly visible to the pilot and unless the airplane is in a position from which a normal landing can be made. (The wisest pilots of all settle only for the approach threshold as a signal to leave the haven of the safe altitude shown on the approach plate.) In fog conditions it is often possible to look straight down and see some portion of the runway when overflying it at the MDA. Or the aircraft might fly into a relatively clear area when it is halfway down the runway. This type of sighting clearly does not satisfy the requirements of the rule.

Strict adherence to the rule on leaving that last altitude would prevent a lot of general-aviation IFR problems, but at Gambell, the MDA rule would have to be paired with something else, because the airplane was only slightly below the MDA at the time of the accident. It was, however, a mile east of the airport, where the terrain is actually higher than the MDA. This points up the necessity of knowing the terrain around an airport and of following the approach procedure to the letter. At Gambell the missed-approach point was the NDB; the missed-approach procedure called for a right turn at that point; and there is a notation that circling east of the airport is not authorized. There is always a good reason for such restrictions on an approach procedure.

Don't think that approaches with such hazards are unique to Alaska. If you turn left instead of right on a missed approach at Williamsport, Pennsylvania, a rocky hillside awaits. It is the same story at Hot Springs, Arkansas: a left turn instead of a right will get you into trouble there. Neither place has radar, so no controller would be able to catch a pilot's error. If there is any weather at all, an instrument approach had best be just that—to the courses flown precisely and minimum altitudes heeded.

Considering the less precise situation, most of us do at times wind up flying marginal VFR in a situation much like the one that existed at Gambell when this accident happened. In such a situation, it is important to remember two principles of survival. The first is related to the V in VFR: visual. There is no way to fly into a rock-strewn cloud if you remain clear of clouds. The other principle is positional awareness. Staying aware of position and of the location of high terrain or obstructions is required to save a fuzzy day. Finally it should be acknowledged that where persistence is an admirable trait in many areas, it isn't necessarily good in flying. The record books are full of accidents that can be attributed to a pilot trying too hard to make a trip, to complete an approach, or to be on time.—R. L. C.

Well—at least they saw the runway. Another airline crew in another part of the world saw nothing at all as they raced along in cloud and darkness, feeling their way at four miles a minute across woods and hills low enough to scare varmints and never knowing it until the last ironic instant. . . —P. G.

25.

RICH MOUNTAIN VOR

Texas International Flight 655, a turboprop Convair 600, left El Dorado, Arkansas at 8:15 P.M. on September 27, 1973, bound for Texarkana. It is a short trip of only 61 nautical miles, and the terrain between the two cities is flat. However, that evening, a line of thunderstorms extended across the route. Some airplanes had penetrated the line, but many had made lengthy detours around the storms. The Convair was equipped with weather radar. The captain of Flight 655 elected not to activate the IFR clearance stored in the computer at Fort Worth Center for his use, planning to penetrate the line of storms VFR. Although uncommon in airline operations, such a plan has a familiar ring in general aviation, despite the fact that night VFR flight in thunderstorm areas has a proven potential for disaster. No contact was made with the Convair after takeoff, and, when it failed to arrive in Texarkana, a search process was initiated. Tragically, an Arkansas National Guard plane crashed during a predawn flight to the search area. Three men died trying to aid fellow airmen in what became the lengthiest search for a U.S. scheduled airliner in memory: 224 aircraft flew over 1,200 search hours; 93 ground vehicles were used. The Convair was found three days after it crashed, 91 miles north of Texarkana. All aboard had perished, but the cockpit voice-recorder was not destroyed. The recording begins a few minutes after takeoff at El Dorado and covers the last 33 minutes and 42 seconds of the flight. There were no communications with ground stations; the entire conversation is between the captain and the first officer, and at first, it centers on radar readings. Time notations are minutes and seconds that have elapsed since the recording began (Capt = Captain; FO = First Officer).

TIME: Shortly after takeoff.
FO—Yeah?
Capt—That might not be a hole there.
FO—We know shortly.
FO—It sorta looks like 24 miles to the end; I don't mind, do you?
Capt—I don't care, just as long as we don't go through it.
FO—Looks a little strange through there; looks like something attenuating through there.
FO—It's a shadow.
Capt—Yeah, looks like a shadow.
Capt—Is that better?
FO—Naw, I don't care.
Capt—Suit yourself.

FO—Well, I don't know, looks a little lighter in here; this thing hits your eyeballs pretty hard.
FO—That's right.
Capt—That's what I'm holding this thing for.
FO—That's all right; it doesn't hurt if she's bright.
Capt—As long as you've got it.
FO—See something?
Capt—I think it's snow.
FO—I still think that's a shadow.
Capt—Yeah, why not.
FO—All right.
Capt—I'd slow it up a little bit, too.
Capt—Good chance.
FO—Whata we got, decreasing ground pickup?
Capt—I didn't hear you.
FO—The visibility is dropping.
Capt—Yeah.
FO—Well.
FO—Rain.
Capt—Raining all over the place.

TIME: 08:40 (into the flight)
Capt—You got something down there?
FO—The other one's a rain data, look at it.
Capt—Yeah, it's probably getting ground clutter down there. That's ground.
FO—What's all this, lights, in those fields? What the—are they chicken farms?
Capt—Yeah.
FO—God Almighty. They're planning on growing a few eggs, ain't they.
Capt—Yep, that's what they are.
  [Whistle.]
  [Sound of ambient noise increases.]
FO—What the—have I got there, ya know?

TIME: 10:18
FO—I thought the end of that line was way back down over there now.

TIME: 12:19
FO—. . .keeps growing on us.
Capt—No, not really.
FO—It's a real cute lil' old curlycue, ain't it?
Capt—Yeah, ha, ha. There's not much to that, but we gotta stay away from it 'cause we'll be vee IFR.
FO—Si—I can't get this—stupid radar—you got any idea where we're at?

TIME: 15:06
Capt—Yeah, 216 will take us right to the VOR.
FO—Two, ah.

154

Capt—Two oh nine, I got.

FO—Fifteen.

Capt—I'm not concerned with that; I could care less. I guess you're right. That, that is just extending on and on and on as we go along 'cause it hasn't moved in about three or four miles in the last 30 minutes, it seems like. I guess it's building up that way now.

[Whistle.]

Capt—What's Hot Springs?

TIME: 16:09

FO—Sir?

Capt—What's Hot Springs VOR? Is it 10:zero; is that right?

FO—Yeah, yeah, uh, that's right. We don't want to get too far up the—it gets hilly.

Capt—Yeah—stars are shining. Why don't you try 2,000?

TIME: 17:03

Capt—If we get up here anywhere near Hot Springs, we get in the—mountains.

FO—Uh, you reckon there's a ridge line along here somewhere? Go down 500 feet; you can see all kindsa lights. Let's go ahead and try for 2,500.

TIME: 17:49

Capt—All right, Fred, you can quit worrying about the mountains 'cause that'll clear everything over there.

FO—That's why I wanted to go to 2,500 feet. That's the Hot Springs highway right here, I think.

Capt—You 'bout right.

FO—Texarkana—naw, it ain't either. Texarkana's back here.

Capt—Texarkana's back over here somewhere.

FO—Yeah. This ain't no Hot Springs highway.

TIME: 18:34

Capt—Well, 30 degrees. . .30 degrees takes you right to Texarkana, doesn't it? Hot Springs—here we are sittin' on 50.

FO—Yeah. How we doin' on the ground?

Capt—I don't know, Fred, still we keep gettin' another one poppin' up every time—every time.

TIME: 19:34

FO—If we keep this up indefinitely, we'll be in Tulsa.

Capt—I haven't been in Tulsa in years.

FO—Ha, ha, ha. The last time, I was with Glen Duke; he said go whichever way you want to. I was going out of Abilene going to Dallas, took up a heading of 010, and flew for about 45 minutes, and he said, "Fred, you can't keep going on this, on this heading." I said, "Why?" He said, "You gonna be in Oklahoma pretty soon." I said, "I don't care if I'm in Oklahoma." He said, "Fair enough."

TIME: 20:14
FO—How'd I get all this—speed?
Capt—You're all right.
FO—Pile it on; we'll keep this speed here?
Capt—A little while.
FO—There ain't no lights on the ground over there.
Capt—Yeah, I see 'em behind us. See stars above us.
FO—I got some lights on the ground.
Capt—There's just not many out here.
FO—Maybe—could be something else, coach.
Capt—Aha, we're gettin' rid of the clouds.
Capt—We is in the clouds, Fred.
FO—Are we?
Capt—Yeah.
Capt—No, we're not.

TIME: 21:16
Capt—I can see above us.
FO—We got ground up ahead?
Capt—I can see the ground here.
FO—Yeah, I can see the ground down here, too.
Capt—Yeah (unintelligible). North is a fair heading. North.
FO—Now, what have we got here?
Capt—Naw, you're all right, I can see some lights over here.
FO—I'll tell ya what, coach. . .

TIME: 22:02
FO—That's probably Hot Springs.
Capt—Yep, could be. Yeah, that might be either it or Arkadelphia.
FO—Well, I'm getting out of the clouds here, Mac, but I'm getting right
straight into it.
Capt—Oh, looks like you're all right.
FO—Do you see any stars above us? We're getting in and out of some scud.
Capt—Yeah. We've got a little bit here.
FO—I sure wish I knew where the—we were.
Capt—Well, I tell you what, we're, ah, on the 250, 260 radial from, ah, Hot
Springs. [Heavy static.]
FO—Figure I can kick her on up here; maybe I can outrun it.
Capt—I don't, I don't think you can get up.
FO—Well, I got the—thing pointed almost straight up to see what we got out
here.

TIME: 23:27
FO—Paintin' ridges and everything else, boss, and I'm not familiar with the
terrain.
FO—We're staying in the clouds.

Capt—Yeah, I'd stay down. You're right in the base of the clouds. I tell you what. We're going be able to turn here in a minute.
FO—You wanna go through there?
Capt—Yeah.

TIME: 24:04
FO—All right. Good, looking good, Mac. Looking good.
Capt—That's all right, wait a minute.
FO—Well, I can't even get, ah, Texarkana anymore.
Capt—I'll tell you what, Fred.
FO—'Kay, boss.
Capt—Well, ah, we'll just try that; we'll try it. We're gonna be in the rain pretty soon; it's only about two miles wide.
FO—You tell me where you want me to go.
Capt—Okay, give me a heading of, ah, 340.
FO—Three-Forty?

TIME: 25:25
Capt—Three-Forty. Steady on. We got 10 miles to go and we're gonna turn—to the left about, ah, about 50 degrees.
FO—Want me to turn—did you say 50?
Capt—Yeah, 50 left. On about, uh, 290.
FO—Two-ninety.
Capt—Ya got six miles to turn.
Capt—Three miles south of turn.
FO—We're in it.
Capt—Huh?
FO—We're in solid, now.
Capt—Are we?
FO—Hold it.

TIME: 27:01
Capt—Start your turn—standard rate—level out and let me see it—when you hit 290.
FO—Aw, okay.
FO—There's your 290.
Capt—Steady on. Should hit in about a half a mile. Should be out of it in about two miles. You're in it. Are you through it?
Capt—Turn 30 left.
FO—I can see the ground now.

TIME: 28:20
FO—There's 30 left. Naw, that's 35.
Capt—Keep on truckin', just keep on a-truckin'.
FO—Well, we must be somewhere in Oklahoma.
Capt—Doin' all the good in the world.
FO—Do you have any idea of what the frequency of the Paris VOR is?

Capt—Nope, don't really give a . . .
Capt—Put, uh, about 265, heading 265.
FO—Heading, 265.
FO—I would say we — up.
Capt—Think so?
  [Laughter.]
Capt—Didn't we?
Capt—Fred, descend to 2,000.

TIME: 30:25
FO—Two-thousand, coming down.
FO—Here we are, we're not out of it.
Capt—Let's truck on.
Capt—About five to the right. Shift over a little bit if you can.
Capt—Sure can. That's all right.
FO—Right.
Capt—That's all right; you're doin' all the good in the world. I thought we'd
get, I thought it was moving that way on me, only we just kinda turned a little
bit while you was looking at the map.
FO—Look.
Capt—First time I've ever made a mistake in my life.
FO—I'll be. Man, I wish I knew where we were, so we'd have some idea of the
general terrain around this place.
Capt—I know what it is.
FO—What?

TIME: 32:01
Capt—That the highest point out here is about 1,200 feet.
FO—That right?
Capt—The whole general area, and then we're not even where that is, I don't
believe.
FO—I'll tell you what, as long as we travel northwest instead of west, I still
can't get Paris.
  [Whistle.]

TIME: 33:18
FO—Two hundred fifty, we're about to pass over Page VOR. You know
where that is?
Capt—Yeah.
FO—All right.
Capt—About 180 degrees to Texarkana.
FO—About 152.

TIME: 33:40
FO—Minimum en-route altitude here is forty-four hund. . .
  [Sound of impact.]

The recording was interrupted when the aircraft hit the steep, heavily wooded north slope of Black Fork Mountain in the Quachita Mountain Range. Impact occurred at an elevation of 2,025 feet, about 600 feet below the top of the ridge. The crash was similar to many other general-aviation and military accidents that have occurred in the area. Those mountains have been hit with regularity since the airplane was invented, including at least two accidents since the Texas International crash. The Page VOR referred to in the transcript even had a broadcast warning on it at the time of the Convair crash, prompted by the large number of accidents: "Page VOR, caution, elevation 2,700 feet." Page VOR has since been renamed Rich Mountain VOR. Hopefully the word "mountain" will suggest something about the terrain to pilots. There is no mysterious force that draws airplanes to dash them against a slope, as some of the local inhabitants have suggested: it is just that mountains are where you find them, not where you think they are. Many pilots think that Arkansas and Oklahoma are flat, but there is much real estate to dodge when flying at 2,000 feet in western Arkansas and eastern Oklahoma. If only the crew of Flight 655 had listened to the identification of Page VOR, they would have heard the terrain warning; if they had used a WAC or sectional chart, as recommended for VFR flight, they would have known about the terrain in the area.

Many people have wondered why the captain elected to go VFR when an IFR clearance was available, and most have strongly questioned his action, but my own experience in flying in the area gives me some insight into why Flight 655 might have set out VFR. The storms that grow in the area tend to generate cloud bases at from 2,000 to 4,000 feet, often making the ride on a low-level VFR flight better than on one at an IFR altitude of about 6,000 feet. The minimum en-route altitude between El Dorado and Texarkana is low enough to allow low-level IFR, but, because the air-traffic-control system often precludes using the lower altitudes, pilots tend to stick to VFR. From the cockpit conversation, it seems that they were not able to maintain VFR, but not knowing the territory over which they were flying during the detour around the storms was the primary cause of the crash. I've talked to a lot of experienced pilots about this accident, and sometimes I've noticed their attention wandering. Their eyes go blank, and a glazed look passes over their faces. Their minds could only be returning to a particular night or day, a cockpit and a line of thunderstorms.—R. L. C.

INDEX OF CONTRIBUTING FACTORS

Numbers after items refer to chapters in which item is treated otherwise than incidentally.